Physiological Noise and Informative Randomness in Neural and Cardiovascular Systems

Andrea Scarciglia • Claudio Bonanno •
Gaetano Valenza

Physiological Noise and Informative Randomness in Neural and Cardiovascular Systems

Springer

Andrea Scarciglia
Neurocardiovascular Intelligence Lab,
Bioengineering and Robotics Research
Centre "E. Piaggio", Department
of Information Engineering
University of Pisa
Pisa, Italy

Claudio Bonanno
Department of Mathematics
University of Pisa
Pisa, Italy

Gaetano Valenza
Neurocardiovascular Intelligence Lab,
Bioengineering and Robotics Research
Centre "E. Piaggio", Department
of Information Engineering
University of Pisa
Pisa, Italy

ISBN 978-3-031-95913-4 ISBN 978-3-031-95914-1 (eBook)
https://doi.org/10.1007/978-3-031-95914-1

This Springer imprint is published by the registered company Springer Nature Switzerland AG
The registered company address is: Gewerbestrasse 11, 6330 Cham, Switzerland

Preface

The concept of noise, intended as stochastic process, takes its roots in 1905, when Einstein *invented* it as a mathematical tool to prove the existence of atoms, revealing also the atom's simple complexity [30]. Before Einstein, scientists like Lavoisier, Dalton, and Avogadro were trying to understand the behavior of chemical reactions through atoms theory, while Clausius, Maxwell, and Boltzmann revolutionized the concept of heat, relating it to the kinetic energy of atoms as formulated in the thermodynamics as we know it today. The latter scientist, in particular, was the first to describe the collisions of atoms in gases introducing stochasticity and providing an equation where the probability density of position and velocity of the atoms evolves through time, irreversibly converging to a Maxwellian distribution. Meanwhile, the Clausius's idea of irreversibility of closed systems in a thermodynamic framework, i.e., the non-decreasing behavior of entropy, started to be interpreted from Boltzmann as a mere statistical number of accessible states of the system.

In his work [37], Einstein predicted a Brownian motion, i.e., the random patterns of particles suspended in a medium (a liquid or a gas), as an observable phenomenon with the aim of proving the existence of atoms, by introducing stochastic components. He derived the governing probability distribution for the small Brownian particles, with a spread proportional to the square root of time, and he indicated how to measure Avogadro's number from a stochastic quantity. In particular, he recognized that the mean kinetic energy of the suspended particles is the same of the mean kinetic energy of the atoms.

Later, the mathematics behind Einstein's idea was simplified and generalized by the physicist Langevin [95], who gave birth to the field of stochastic differential equation becoming the first in history to use Newton's equations with a random force. In particular, he derived Einstein's equation for the description of the Brownian motion starting from the deterministic Newton's equations of a ball going through a fluid with a friction—the Stokes law—and adding another irregular, stochastic force due to the collisions of the surrounding molecules. Einstein's idea was verified by Perrin who also was able to compute Avogadro's number.

Other scientists like Smoluchowski, Ornstein, Rice, Wiener, Kolmogorov, Itô, and Stratonovich have improved the mathematics of noisy processes, allowing the formal definition of differential equations contaminated by noise—it is possible to define processes whose derivative is pure noise or the integral of noise patterns, and a rigorous description of the Brownian motion—and methods to solve these (e.g., the Langevin's or Fokker-Plank's approaches). Of note, stochasticity has been also introduced in financial engineering by Bachelier in the early years of twentieth century. His aim was to observe how stock prices changed in time from a probabilistic viewpoint [8].

Noise has been also important in electrical engineering for many aspects: for the development and the understanding of the vacuum tube; for the modulation theory, i.e., the techniques which allow the transmission of the radio signals through amplitude or frequency modulation, since the methods have been improved to be less sensitive to the presence of additive noise; in communication theory, since probability and uncertainty play an important role into information transmission. In these fields, the main contributors were Schottky, Johnson, Nyquist, and Shannon. Schottky, moreover, considered and defined what is called shot noise and thermal noise: the former is modeled with Poissonian processes and it is associated with the discrete nature of the electric charge; the latter, instead, refers to the loss of some electrons in a conductor. Noise propagation and transformation through electrical devices has been also of great interest for engineers like Rice who provided the foundation and methodology for modern noise calculations, proving fundamental results and devising method for noise transformation by electrical devices [136].

In addition, it is worthwhile to mention that the study of stochastic processes allowed to recover propagated signals embedded in noise through the Middleton's and North's concept of the matched filter.

Noise is also important in astronomy because of its ubiquitously presence in the universe, which disproved the theory of the "steady state" universe (i.e., it had the same form from its origin) in favor of the acknowledged "Big Bang" theory. Moreover, noise plays a fundamental role in spectroscopy since it modifies the widths of the spectral lines defining the environmental parameters, such as temperature and pressure of the several atoms or molecules composing the universe.

Beyond its foundational role in theoretical science and engineering, noise has recently attracted growing interest in the field of physiology. Increasing evidence suggests that noise is not merely a mere disturbance but may play a crucial role in shaping and even enabling key internal functions of biological systems. With this perspective in mind, in this book we set out to provide readers with a comprehensive overview on the role of stochasticity in neural and cardiovascular systems—commonly referred to as physiological noise. We emphasize the use of a hybrid analytical framework that integrates concepts from both probability theory and dynamical systems—disciplines that traditionally view systems through opposing lenses: one treating randomness as fundamental, the other emphasizing deterministic evolution. Within this framework, we analyze noise—modeled as a stochastic process—by leveraging the sensitivity of a nonlinear complexity measure, Approximate Entropy (ApEn) [132], originally developed to assess regularity in dynamical systems. This allows us to detect and quantify stochastic contributions embedded within physiological dynamics.

In this context, each physiological system is conceptualized as a dynamic process governed by an unknown deterministic map, with noise dynamically superimposed. This approach is applied to univariate (e.g., brain or cardiac dynamics) physiological systems, enabling the estimation of noise levels without requiring explicit knowledge of the underlying analytical equations. Furthermore, the framework supports the evaluation of noise non-stationarity across time, offering insights into how stochastic properties evolve in physiological systems. In cases where the dynamics are sufficiently regular, the method also enables the reduction—or even partial removal—of the noise component, revealing clearer patterns of underlying activity.

Pisa, Italy

Andrea Scarciglia
Claudio Bonanno
Gaetano Valenza

Declarations

Competing Interests The authors have no conflicts of interest to declare that are relevant to the content of this book.

Contents

List of Abbreviations

ANS	Autonomic Nervous System. 37
ApEn	Approximate Entropy. 20
AR	Autoregressive Models. 62
CDF	Cumulative Distribution Function. 3
CDF	Probability Density Function. 3
CHM	Cumulative Histogram Method. 57
CNS	Central Nervous System. 37
ECG or EKG	Electrocardiogram. 39
EEG	Electroencephalography. 49
fMRI	Functional Magnetic Resonance Imaging. 50
HRV	Heart Rate Variability. 39
IID	Independent and Identically Distributed. 11
MSE	Mean Squared Error. 65
PMF	Probability Mass Function. 4
PNS	Parasympathetic Nervous System. 37
PPG	Photoplethysmogram. 39
PSD	Power Spectral Density. 23
SNS	Sympathetic Nervous System. 37

Notation

(X, T)	Dynamical System desctibed the action of the Map T on the space X. 11
(X, T, μ)	Metrical and Measure-Preserving Dynamical System. 13
$(\Omega, \mathcal{F}, P)$	Probability Space. 2
$C(r)$	Correlation Integral. 20
C^1	Space of the Differentiable Maps or Functions. 54
D_B	Box-Counting Dimension. 19
D_C	Correlation Dimension. 20
D_H	Hausdorff Dimension. 19
F_Y	Cumulative Distribution Function of a Random Variable Y. 3
T	Map, usually continuous transformation, which defines the dynamical system on a given phase space. 11
X	Phase Space on which is deined a dynamical system. 11
Y	Random Variable. 2
η_Y or η	Expected Value of a Random Variable Y. 6
$\\|\cdot\\|$	Infinity or Maximum norm in $\mathbb{R}^m$. 60
$\lambda(x, v)$	Lyapunov Exponent of an orbit starting from the point x. 17
$\{\varepsilon_n\}_{n\geq 0}$	Realization of a Random Variable, usually supposed Gaussian or Normal.. 28
$\{x_n\}_{n\geq 0}$	Trajectory or Orbit of a Metrical Dynamical System.. 28
$\mathbb{E}[Y]$	Expected Value of a Random Variable Y. 6
$\mathbf{Y}$	Set of Random Variables. 6
$\mathcal{F}$	σ-algebra. 2
$\mathcal{N}(\eta, \sigma^2)$	Gaussian or Normal Random Variable with median η and variance σ^2. 7
μ	Probability Measure. 2
σ	Standard Deviation of a Gaussian or Normal Random Variable, often denoting the level of estimated noise witgin the series. 7
KS	Kolmogorov-Sinai Entropy. 17
Var(Y)	Variance a Random Variable Y. 6
std	Standard Deviation of a Random Variable Y. 6

Ω	Space of Events. 2
Σ	Covariance Matrix of a Set of Random Variables. 7
f_Y	Probability Density Function of a Random Variable Y. 3
m	Embedding Dimension of a reconstructed phase space . 21
r	Radius or Tolerance which defines the kernel amplitudes of the ApEn or SampEn quantifiers . 20

List of Figures

List of Tables

Chapter 1
Introduction and Key Definitions

Abstract This chapter introduces the two theoretical frameworks employed to analyze noise in physiological signals: probability theory and dynamical systems theory.

The first framework—probability theory—revisits fundamental concepts such as random variables and probability distributions, along with their key properties. Particular attention is given to stochastic processes, which are essential for modeling random phenomena. Within this context, the focus is on stationary processes, defined as those whose statistical characteristics (e.g., mean and variance) remain constant over time.

The second framework—dynamical systems theory—addresses the deterministic aspects of systems evolving in metric spaces equipped with probability measures. Key features of deterministic dynamics are introduced, including sensitive dependence on initial conditions, chaos, and the concept of attractors. The chapter also outlines the principal quantifiers used in the analysis of dynamical systems, with a particular emphasis on Kolmogorov–Sinai (K–S) entropy and its practical approximation for finite time series, namely Approximate Entropy (ApEn).

Keywords Probability · Stochastic processes · Dynamical systems

1.1 Concepts of Probability Theory

Probability theory is the branch of mathematics concerned with the analysis of random phenomena, properly analyzed through the *probability space* objects, which involve the probability measure on a given set Ω and the σ-algebra concepts. Intuitively, the former consists in a rigorous way to quantify the likelihood of events occurring in a set Ω, while the latter defines all the possible sub-events of Ω on which it is possible to express the occurrence likelihood. Formally, one has the following definition:

A. Scarciglia et al., *Physiological Noise and Informative Randomness in Neural and Cardiovascular Systems*, https://doi.org/10.1007/978-3-031-95914-1_1

Definition 1.1.1 (Probability Measure) A measure $\mu : \mathcal{F} \to \mathbb{R}$ on a σ-algebra $\mathcal{F}$ on space Ω is said of probability if:

- μ takes values on $[0, 1]$.
- $\mu(\Omega) = 1$ and $\mu(\emptyset) = 0$.
- It is countable additive, i.e., for pairwise disjoint sets $S_i \in \mathcal{F}$ with $i \in \mathbb{N}$, it holds $\mu(\cup_{i\geq 0} S_i) = \sum_{i\geq 0} \mu(S_i)$

Definition 1.1.2 (σ-algebra) A σ-algebra $\mathcal{F}$ on a set Ω is a non-empty collection of subsets of Ω, which is:

- Closed under complement, i.e., if $S \in \mathcal{F}$, then $X/S \in \mathcal{F}$
- Closed under countable unions, i.e., if $S_i \subset \mathcal{F}\ \forall i$, then $\cup_{i\geq 0} S_i \in \mathcal{F}$
- Countable intersections, i.e., if $S_i \subset \mathcal{F}\ \forall i$, then $\cap_{i\geq 0} S_i \in \mathcal{F}$

Now, it is possible to define the probability space framework.

Definition 1.1.3 (Probability Space) A probability space is defined as a triple $(\Omega, \mathcal{F}, P)$, where:

- Ω is the sample space, representing all possible outcomes of a random experiment.
- $\mathcal{F}$ is a σ-algebra of subsets of Ω, representing events that can occur.
- P is a probability measure on $\mathcal{F}$.

Now, some of the fundamental concepts of probability theory have been reviewed. These include random variables, probability distributions, cumulative distribution functions, probability densities, mean, and covariance.

The random variable is a mapping which allows to work with probabilities in a numerical way, making it easier to analyze and model random phenomena.

Definition 1.1.4 (Random Variable) A random variable (r.v.) Y is a measurable function from the sample space Ω to the real numbers $\mathbb{R}$ (or in general to a measurable space).

It is possible to define the likelihood of outcomes of a random variable Y, considering the probability P that Y takes on a value in a measurable set $S \subseteq \mathbb{R}$ as

$$P(Y \in S) = \mu(\{\omega \in \Omega : Y(\omega) \in S\}). \tag{1.1}$$

The probability distribution is intuitively a complete description of the probabilities of the possible values of a random variable Y, showing how likely it is for a random variable to take on each of its possible values. It provides a mapping of the probabilities associated with all potential outcomes of a random experiment.

Definition 1.1.5 (Probability Distribution) The probability distribution of a random variable Y is the collection of all the probabilities $P(Y = x)$, with $x \in \mathbb{R}$.

In particular, the probability distribution does not take into account the specific probability space used to define Y and solely records the probabilities of the different output values of Y.

The random variables can be roughly categorized into *discrete* if they take values on a finite or a countable subset of $\mathbb{R}$ and *continuous* if the codomain is continuous. In the former, the binomial, Poisson, Geometric, and Hypergeometric distributions are found; the latter family, instead, comprises the normal or Gaussian, the uniform, exponential, Gamma, Beta, and Rice distributions.

For a real-valued random variable Y—either continuous or discrete—its probability distribution can be represented by the following definition:

Definition 1.1.6 (Cumulative Distribution Function (CDF)) For a random variable Y with probability distribution P, the Cumulative Distribution Function (CDF) F_Y is defined as

$$F_Y(x) = P(Y \leq x) \tag{1.2}$$

for any x in the sample space.

The CDF F_Y is non-decreasing as function of $x \in \mathbb{R}$, and it approaches to 0 and to 1 when, respectively, $x \to +\infty$ and $x \to -\infty$, i.e.,

$$\lim_{x \to -\infty} F_Y(x) = 0 \text{ and } \lim_{x \to +\infty} F_Y(x) = 1.$$

The CDF accumulates the probabilities assumed by the outcomes of the random variables up to a specific point and gives their total probability of being less than or equal to a specific value.

Under certain assumptions of regularity, it can be derived the probability density function from the CDF of a continuous random variable.

Definition 1.1.7 (Probability Density Function (PDF)) For a random variable Y with CDF F_Y, the Probability Density Function (PDF) f_Y is defined as the derivative of the CDF

$$f_Y(x) = \frac{d}{dx} F_Y(x)$$

for any x in the sample space, provided that the derivative exists.

The PDF f_Y is nonnegative ($f(x) \geq 0$ for all x) and standardized ($\int_{-\infty}^{\infty} f_Y(x)dx = 1$) and describes the probability that a random variable Y falls within the interval $[a, b]$ according to

$$P(a \leq Y \leq b) = \int_a^b f(x)dx.$$

The PDF describes the relative likelihood of a continuous random variable taking on a specific value, providing the probability density at each point in the variable's range, which means that the probability of the variable falling within a certain interval can be found by integrating the PDF over that interval.

If the r.v. is discrete, alternatively to the PDF, it can be defined the Probability Mass Function (PMF) p, which assigns a probability to each possible discrete outcome x, such that

$$p(x) = P(Y = x) \text{ and } \sum_x p(x) = 1.$$

In probability theory one has to analyze the interactions among different random variables defined on the same event space. Hence, the useful previous definitions of CDF and PDF can be extended to a set of random variables $Y_1, \ldots, Y_N$, by considering the probabilities that all the random variables take on specific outcomes. One can find the following definition:

Definition 1.1.8 (Joint CDF) Given N random variables $Y_1, \ldots, Y_N$, the joint cumulative distribution function (joint CDF) $F_{Y_1,\ldots,Y_N}$ is defined as the probability P that the random variable Y_1 takes on a value less than or equal to x_1, Y_2 takes on a value less than or equal to x_2, and so on up to Y_n that takes on a value less than or equal to x_n, in symbols

$$F_{Y_1,\ldots,Y_N}(x_1, \ldots, x_N) = P(Y_1 \leq x_1, \ldots, Y_N \leq x_N) \tag{1.3}$$

for any samples $x_i \in Y_i, \forall i = 1 \ldots N$.

Analogously, one has the following definition:

Definition 1.1.9 (Joint PDF/PMF) In the same setting of the previous definition, if the N random variables are continuous, then it is possible to define the joint probability density function as $f_{Y_1,\ldots,Y_N}(x_1, \ldots, x_N)$

$$f_{Y_1,\ldots,Y_N}(x_1, \ldots, x_N) = \frac{\partial^n F_{Y_1,\ldots,Y_N}(x_1, \ldots, x_N)}{\partial x_1 \partial x_2 \ldots \partial x_n}, \tag{1.4}$$

where $F_{Y_1,\ldots,Y_N}(x_1, \ldots, x_N)$ is the joint CDF of Eq. (1.3), provided that the derivative exists. Alternatively, if the N random variables are discrete, then the joint mass function $p_{Y_1,\ldots,Y_N}(x_1, \ldots, x_N)$ is defined as

$$p_{Y_1,\ldots,Y_N}(x_1, \ldots, x_N) = P(Y_1 = x_1 \text{ and } \ldots \text{ and } Y_N = x_N). \tag{1.5}$$

By denoting the random vectors $\mathbf{Y} = (Y_1, \ldots, Y_N)^{'}$ and $\mathbf{x} = ((x_1, \ldots, x_N)^{'})$ ($'$ denotes the transposition operation), then it is possible to use a shorter notation for the joint CDF $F_{\mathbf{Y}}(\mathbf{x})$, for the joint probability mass function $p_{\mathbf{Y}}(\mathbf{x})$, and for the joint probability density function $f_{\mathbf{Y}}(\mathbf{x})$.

From the joint CDF and PDF it is possible to retrieve the CDF and PDF of a single random variable, called *marginal* CDF and PDF.

Definition 1.1.10 (Marginal CDF) Given N random variables $Y_1, Y_2, \ldots, Y_N$ and their joint CDF $F_{Y_1,\ldots,Y_N}(x_1, \ldots, x_N)$, the marginal CDF of the random variable Y_1 is given by

$$F_{Y_1}(x_1) = \lim_{x_2 \to \infty} \cdots \lim_{x_N \to \infty} F_{Y_1,\ldots,Y_N}(x_1, \ldots, x_N).$$

The marginal CDF of the other random variables $F_{Y_i}(x_i)$ for $i = 2, \ldots, N$ is obtained by fixing the i-th sample of the joint CDF and sending to infinity the remaining variables.

Definition 1.1.11 (Marginal PDF and PMF) Given N continuous random variables $Y_1, Y_2, \ldots, Y_N$ and their joint PDF $f_{Y_1,\ldots,Y_N}(x_1, \ldots, x_N)$, the marginal PDF of the random variable Y_1 is given by

$$f_{Y_1}(x_1) = \int_{-\infty}^{\infty} \cdots \int_{-\infty}^{\infty} f_{Y_1,\ldots,Y_N}(x_1, \ldots, x_N) dx_2 \cdots dx_N.$$

The marginal PDF of the other random variables $F_{Y_i}(x_i)$ for $i = 2, \ldots, N$ is obtained by fixing the i-th sample of the joint CDF and by integrating the remaining variable in their domain.

If the N variables are discrete, then the marginal PMF of the random variable Y_1 is given by

$$p_{Y_1}(x_1) = \sum_{x_2=-\infty}^{\infty} \cdots \sum_{x_N=-\infty}^{\infty} p_{Y_1,\ldots,Y_N}(x_1, \ldots, x_N).$$

The marginal PMF of the other random variables $p_{Y_i}(x_i)$ for $i = 2, \ldots, N$ is obtained by fixing the i-th sample of the joint PMF and by integrating the remaining variable in their domain.

The joint CDF or PDF allows us to establish the *independence* of random variables, which is a fundamental concept in probability theory. Two (or more) random variables are said to be independent if knowing the value of one random variable provides no information about the value of the other. This means that the occurrence of one event does not influence the occurrence of the other. Formally, one has the following definition:

Definition 1.1.12 (Mutual Independence of Random Variables) Given N random variables $Y_1, Y_2, \ldots, Y_N$, these are mutually independent if, for any subset of these variables and for all values that the variables can take, the joint CDF of the subset can be factorized into the product of their marginal distributions

$$F_{Y_1,\ldots,Y_N}(x_1, \ldots, x_N) = F_{Y_1}(x_1), \ldots, F_{Y_N}(x_N). \tag{1.6}$$

This factorization easily extends to the joint PDF and the joint PMF.

Three important features of the random variables are the expected value, the variance, and the standard deviation. The former provides a measure of the central tendency of a random variable's distribution, consisting in a weighted average, where each possible value of the random variable is weighted by its probability. The variance, instead, quantifies the dispersion or variability of a random variable's distribution, analogously to the related standard deviation, which in turn uses the same units of the original data.

Definition 1.1.13 (Expected Value or Mean) The expected value (or mean) of a random variable Y, denoted $\mathbb{E}[Y]$ or η_Y or η when no ambiguity is present, is a measure of the central tendency of the distribution of Y. For continuous r.v., it is defined as

$$\mathbb{E}[Y] = \int_{-\infty}^{\infty} (1 - F_Y(x))dx \text{ (using CDF)}$$

$$\mathbb{E}[Y] = \int_{-\infty}^{\infty} (xf_Y(x))dx \text{ (using PDF).}$$

For discrete random variables, instead, it is defined as

$$\mathbb{E}[Y] = \sum_x xp(x) \text{ using the PMF}$$

for discrete random variables.

Analogously, for a finite set of random variables $\mathbf{Y} = (Y_1, \ldots, Y_N)^{'}$, the expected value $\mathbb{E}[\mathbf{Y}]$ is the vector composed of the mean of the random variables, i.e.,

$$\mathbb{E}[\mathbf{Y}] = (\mathbb{E}(Y_1), \ldots, \mathbb{E}(Y_N))^{'}.$$

Definition 1.1.14 (Variance) The variance of a random variable Y, denoted $\text{Var}(Y)$, measures the spread or dispersion of the values of Y around its mean $\mu = \mathbb{E}[Y]$. If the random variable is continuous, then

$$\text{Var}(Y) = \mathbb{E}[(Y - \eta)^2] = \int_{-\infty}^{\infty} x(x - \eta)^2 df_Y(x).$$

If X is discrete, then

$$\text{Var}(Y) = \mathbb{E}[(Y - \eta)^2] = \sum_x x(x - \eta)^2 p(x). \tag{1.7}$$

Definition 1.1.15 (Standard Deviation) The standard deviation (std) of a random variable X is the square root of its variance, namely $\text{std}(Y) = \sqrt{\text{Var}(Y)}$.

While the expected values can be easily extended to a set of random variables, the variance is extended to multiple random variables in terms of statistical covariance, which measures the degree to which two (or more) random variables change together.

Definition 1.1.16 (Covariance of Random Variables) For a set of random variables $\mathbf{Y}$, the covariance is described by the matrix Σ, defined as

$$\Sigma = \mathrm{Cov}(\mathbf{Y}) = \mathbb{E}[(\mathbf{Y} - \boldsymbol{\eta})(\mathbf{Y} - \boldsymbol{\eta})^{'}],$$

where the entries of Σ are

$$\Sigma_{ij} = \mathrm{Cov}(Y_i, Y_j) = \mathbb{E}[(Y_i - \mathbb{E}[Y_i])(Y_j - \mathbb{E}[Y_j])].$$

1.1.1 Normal or Gaussian Distribution

Here a peculiar probability distribution is presented. This is the normal distribution, also known as the Gaussian distribution, which is a continuous probability distribution characterized by its symmetric bell-shaped curve. It is widely used in natural sciences to model real-valued random variables whose distributions are not known. The normal distribution is fundamental in statistics as well, since many statistical tests for hypothesis testing exploit normal distribution properties in inferential statistics.

The PDF of a normal distribution is given by the formula

$$f(x) = \frac{1}{\sigma\sqrt{2\pi}} e^{-\frac{(x-\eta)^2}{2\sigma^2}}, \tag{1.8}$$

where:

- η is the mean of the distribution.
- σ is the standard deviation.
- e is Euler's number (approximately 2.71828).

Denote by $Y \in \mathcal{N}(\eta, \sigma^2)$ a random variable X which has Gaussian or normal with median η and variance σ^2.

Some properties can be summarized here:

- The curve is symmetric about the mean η.
- Approximately 68.27% of the data falls within one standard deviation ($\eta \pm \sigma$).
- Approximately 95.45% of the data falls within two standard deviations ($\eta \pm 2\sigma$).
- Approximately 99.73% of the data falls within three standard deviations ($\eta \pm 3\sigma$).

The CDF of a normal distribution is given by

$$F(x) = P(Y \leq x) = \frac{1}{2}\left(1 + \operatorname{erf}\left(\frac{x-\eta}{\sigma\sqrt{2}}\right)\right), \tag{1.9}$$

where erf is the error function

$$\operatorname{erf}(x) = \frac{2}{\sqrt{\pi}} \int_0^x e^{-t^2}\, dt$$

corresponding to an integral of the exponential function e^{-t^2}.

1.2 Stochastic Processes

In this section, mathematical methods for describing the systems evolving over time, which present uncertainty or randomness, are presented. These objects are called *stochastic or random processes* and are indeed used for modeling phenomena where the future states are not fully determined by its initial state—like the deterministic processes of the next section—but are influenced by random, uncertain factors.

The stochastic processes are essentially a collection of random variables indexed by time.

Definition 1.2.1 (Stochastic or Random Process) A stochastic or random process is a collection of random variables $Y(\omega, t) : \Omega \times \mathcal{I} \to \mathbb{R}$ defined in the same probability space $(\Omega, \mathcal{F}, \mu)$, letting t vary in the index set $\mathcal{I}$.

Again, Ω is the space of realizations, $\mathcal{F}$ is the associated σ-algebra, and μ is the probability measure defined on the event space Ω. Hence, the random process is a two-variable function, leading to the following interpretations:

- For a fixed $t^* \in \mathcal{I}$, $Y(\omega, t^*)$ is a random variable.
- For a given outcome ω^*, $Y(\omega^*, t) : T \to \mathbb{R}$ means that as the index t varies, the probability associated with the same outcome ω^* is changing, because the random variable itself is varying. In other words, for a given ω^*, the function $Y(\cdot, \omega^*) : \mathcal{I} \to \mathbb{R}$ defines a *sample function* (also known as *realization*, *sample path*, or *trajectory*) of the stochastic process $\{Y(t, \omega)\} : \mathcal{I} \to \mathbb{R}$.

For the following discussion, it is convenient to drop in the notations explicit reference to the dependence on ω and indicate the stochastic process as $Y(t)$ with $t \in \mathcal{I}$. The index set $\mathcal{I}$ is usually referred to time, but it can also be an index referring to space.

If $\mathcal{I}$ is a continuous subset of $\mathbb{R}$, then the random process is said *continuous-time*, and if $\mathcal{I}$ is finite or countable (e.g., $\mathcal{I} = \mathbb{N}$ or $\mathbb{Z}$), the process is *discrete-time*. Analogously, if the random variables $Y(t)$ take discrete or finite values in $\mathbb{R}$, $\forall t \in \mathcal{I}$, then the random process is said *discrete-space*; otherwise, it is called *continuous-space*.

From the previous definition, it follows that each of the random variables composing a stochastic process could not follow the same probability distribution. Indeed, at each point $t \in \mathcal{I}$, the random variable $Y(t)$ follows a certain probability distribution, changing accordingly to t. For this reason, it is convenient to study the stochastic processes in terms of the *joint CDF* $P_{Y(t_1),Y(t_2),\ldots,Y(t_N)}(x_1, x_2, \ldots, x_N)$ for any N finite indexes $t_1, \ldots, t_n \in \mathcal{I}$.

When analyzing stochastic processes, it can be important to establish the degree to which the values of a process at one time are linearly related to its values at another time, for periodicity and recurrent patterns identification. In this direction, a useful tool is the *autocovariance*, which exactly measures the covariance of the process with itself at two different time points. Formally, one has the following definition:

Definition 1.2.2 (Autocovariance) For a stochastic process $\{X(t) : t \in \mathcal{I}\}$, the autocovariance function between time points t and s is defined as

$$\gamma(t, s) = \mathrm{Cov}(Y(t), Y(s)) = E\left[(Y(t) - E[Y(t)])(Y(s) - E[Y(s)])\right].$$

Alternatively, it can be expressed as

$$\gamma(t, s) = E[Y(t)Y(s)] - E[Y(t)]E[Y(s)].$$

Next, we consider three key properties of stochastic processes: strict-sense stationarity, weak-sense stationarity, and ergodicity. Intuitively, stationarity refers to the invariance of a process's statistical properties—such as mean, variance, and autocovariance—over time. Varying degrees of this invariance give rise to different types of stationarity.

Strict-sense stationarity, also known as strong stationarity, requires that all statistical properties of the process remain constant over time. Formally, this condition is satisfied when the joint probability distributions of the process are invariant under time shifts. Formally, one has the following definition:

Definition 1.2.3 (Strict-Sense Stationarity) A process is strict-sense stationary if the joint probability distribution satisfies

$$\begin{aligned} &P(Y(t_1) \le x_1, Y(t_2) \le x_2, \ldots, Y(t_n) \le x_n) = \\ &\quad P(Y(t_1 + \tau) \le x_1, Y(t_2 + \tau) \le x_2, \ldots, Y(t_n + \tau) \le x_n) \end{aligned}$$

for all admissible $\tau \in \mathbb{R}$ and time points $t_1, t_2, \ldots, t_n \in \mathcal{I}$.

Strict-sense stationarity condition is often too restricting for random processes analysis because of the demanding task of retrieving the joint cumulative distribution function from the observed available sample paths, which might not comprise all the possible outputs. For this reason, stationarity condition is weakened requiring only that mean and autocovariance function of the process do not depend on a specific time. This is called wide-sense stationarity.

Definition 1.2.4 (Weak Stationarity) A process is wide-sense stationary (WSS) if its mean and autocovariance function depend only on the time difference (lag) τ, not on the specific time t. That is,

$$E[Y(t)] = \eta \quad \forall t \in \mathcal{I},$$

and the autocovariance function satisfies

$$\mathrm{Cov}(Y(t), Y(t+\tau)) = \gamma(\tau) \quad \forall t \in \mathcal{I},$$

and for all admissible $\tau \in \mathbb{R}$, where $\gamma(\tau)$ is the autocovariance function that depends only on the lag τ.

It can be shown that strict stationarity implies wide-sense stationarity—the converse is not true. In spite of this, wide-sense stationarity is simpler to prove with respect to strict-sense stationarity.

In applied sciences, while wide-sense stationarity is often assumed for modeling phenomena, the problem of characterizing mean and autocovariance remains. Indeed, in order to accomplish this, multiple realizations or sample path of the same phenomenon are required, which in most cases are not available.

To overcome this issue, the *ergodicity* is often assumed for random processes. Briefly, this property states that is possible to retrieve statistical properties using uniquely a long observed sample path. The main ergodic properties are

Definition 1.2.5 (Ergodicity in the Mean) Let $Y(t)$ be a wide-sense time-discrete stationary process with mean η and autocovariance $\gamma_Y(t_1, t_2) = \gamma_Y(t_1 - t_2) = \gamma(\tau)$. The process $Y(t)$ is ergodic in the mean if the time average of its realizations $\bar{x}_N = \frac{1}{N}\sum_{t=1}^{N} x_t$ converges in quadratic mean to η as N increases to ∞,

$$\lim_{N \to \infty} \mathbb{E}[(\bar{x}_N - \eta)^2] = 0. \tag{1.10}$$

Analogously, for a continuous-time WSS stochastic process $Y(t)$, ergodicity in the mean implies that the mean $\bar{x}_T = \frac{1}{T}\int_{t=1}^{T} x(t)dt$ converges in quadratic mean to η as T increases to ∞.

Definition 1.2.6 (Ergodicity in the Covariance) Let $Y(t)$ be a wide-sense time-discrete stationary process with mean η and autocovariance $\gamma_Y(t_1, t_2) = \gamma_Y(t_1 - t_2) = \gamma(\tau)$. The process $Y(t)$ is ergodic in the covariance if the time average of its realizations $\bar{\gamma}_N(\tau) = \frac{1}{N}\sum_{t=\tau+1}^{N}(x_t - \bar{x}_N)(x_{t-\tau} - \bar{x}_N)$ for $\tau = 0, 1, \ldots, N-1$, converges in quadratic mean to $\gamma(\tau)$ as N increases to ∞,

$$\lim_{N\to\infty} \mathbb{E}[(\bar{\gamma}_N(\tau) - \gamma(\tau))^2] = 0. \tag{1.11}$$

Analogously, for continuous-time WSS stochastic process $Y(t)$, ergodicity in the covariance implies that the mean $\bar{\gamma}_T(\tau) = \frac{1}{T}\int_{t=\tau+1}^{T}(x(t) - \bar{x}_T)(x(t-\tau) - \bar{x}_T)dt$ converges in quadratic mean to $\gamma(\tau)$ as T increases to ∞, for any admissible τ.

To resume, the ergodic property in the mean (or in the covariance) states that it is possible to determine the statistical properties mean and covariance of the process from averaging the process samples instead of the different realizations for a fixed time. In ergodicity, moreover, any single trajectory or realization of the process is representative of the entire statistical behavior of the process. Some properties of ergodicity are reported below:

- Ergodicity in covariance implies ergodicity in mean. The converse is not true.
- Ergodicity in mean or in variance does imply wide-sense stationarity; the converse is not true since one can have wide-sense stationary processes which are not ergodic.
- The ergodic property can be generalized to any moment of the process.

To conclude, another important class of stochastic processes is formed by the *independent and identically distributed* random variables, in which the outcome of one random variable does not influence the outcome of any other and all the random variables in the sequence have the same probability distribution, following the same pattern of probabilities.

Definition 1.2.7 (IID Processes) The independent and identically distributed (IID) random processes are stochastic processes whose components are formed by identically distributed random variables and that are mutually independent.

It can be shown that IID processes are wide-sense stationary and ergodic in the mean and in the covariance. This assumption allows us to retrieve the mean and covariance of the process by only one single realization of the process.

1.3 Dynamical Systems Theory

Dynamical systems theory is a branch of mathematics concerned with describing the time-dependent and long-term behavior of evolving systems. In the broadest sense, dynamical systems can be classified as either discrete-time or continuous-time, depending on whether their evolution is observed at distinct time intervals or over a continuous time domain.

Definition 1.3.1 A discrete-time dynamical system is a couple (X, T) representing the action of a map $T : X \to X$ on a non-empty set X, called *state space*, for which it is possible to define the $n-th$ iterate $T^n = T \circ T \circ \cdots T$ (n times) for any positive integer $n \in \mathbb{N}$, being T^0 the identity map, denoted as Id. Since $T^{n+m} = T^n \circ T^m$, the iterates form a *semigroup*. If T is invertible, then $T^{-n} = T^{-1} \circ T^{-1} \circ \cdots T^{-1}$ n times, and the iterates $\{T^q | q \in \mathbb{Z}\}$ form a group through the composition operator.

Definition 1.3.2 A continuous-time dynamical system is a couple (X, T) consisting of a non-empty *state space* X and of a one-parameter family of maps T^t, with $t \in \mathbb{R}^+ \cup \{0\}$, that forms a one-parameter semigroup, i.e., $T^{t+s} = T^t \circ T^s$ and $T^0 = \text{Id}$. The resulting dynamical system is called a semiflow.

Instead, if $t \in \mathbb{R}$, then T^t is invertible, with iterates forming a group, and the resulting dynamical system is called a flow.

Unless otherwise specified, the focus throughout this book is on discrete-time dynamical systems. This choice reflects the nature of physiological data, which is typically acquired as discrete-time series or inherently represents discrete-time processes. While many theoretical concepts and results are applicable to both discrete-time and continuous-time systems, the discrete-time perspective is emphasized here, with the understanding that continuous-time analogs can often be inferred accordingly.

Within dynamical systems theory, particular importance is placed on orbits—the trajectories traced by points in the system's state space as they evolve over time according to the governing rules of the system. More specifically, one has the following definition:

Definition 1.3.3 Let t range over $\mathbb{N} \cup \{0\}$. For $x \in X$, the *positive semiorbit* $\mathcal{O}_T^+(x)$ is defined as the subset of the state space X formed by positive iterates of x

$$\mathcal{O}_T^+(x) = \cup_{t \geq 0} T^t(x).$$

Analogously, if t ranges over $\mathbb{Z}$, one can define the *negative semiorbit* $\mathcal{O}_T^-(x)$ as

$$\mathcal{O}_T^-(x) = \cup_{t \leq 0} T^t(x),$$

and the *orbit* $\mathcal{O}_T(x)$ as

$$\mathcal{O}_T(x) = \mathcal{O}_T^+(x) \cup \mathcal{O}_T^-(x)$$

for any $x \in X$.

The previous definition applies to continuous-time dynamical systems if t varies over $\mathbb{R}$.

Although the general abstract framework is introduced for dynamical systems, concrete applications often involve systems with specific underlying structures. For example, a dynamical system might be defined on a measure space with a measure-preserving transformation, on a topological space with a continuous map, on a metric space with an isometry, or on a smooth manifold with a differentiable map.

For the following discussion, dynamical systems are equipped with metrics and probability measures.

Definition 1.3.4 A metrical and measure-preserving dynamical system (X, T, μ) is a dynamical system (X, T, μ) in which:

- A distance $d : X \times X \to \mathbb{R}$ is defined.
- T is a continuous function.
- $\mu : \mathcal{F} \to [0, 1]$ is a probability measure on the space X.
- μ is preserved by the map measurable T transformation, i.e., $\forall A \in \mathcal{F}$ it holds $\mu(T^{-1}(A)) = \mu(A)$.

Measure-preserving dynamical systems are the fundamental concept in dynamical systems theory, especially in ergodic theory. These systems, which adhere to the Poincaré recurrence theorem, are a specific type of conservative system. They serve as the mathematical foundation for a wide range of physical systems, including many from classical mechanics (particularly non-dissipative ones) and those in thermodynamic equilibrium.

Among all the dynamical system features, it has been given attention on the sensitive dependence on initial conditions, on chaotic traits, and on the determination of the attractor properties of a dynamical system. The feature of sensitive dependence on initial conditions refers to the fact that small variations in the initial state can result in drastically different long-term behavior, even though the governing equation is deterministic. Formally, one has the following definition:

Definition 1.3.5 (Sensitive Dependence on Initial Conditions) A dynamical system (X, T) has sensitive dependence on initial conditions [22, 165] on $S \subset X$ if there is $r > 0$, such that for every $x \in S$ and $\delta > 0$ there are $y \in X$ and $n \in \mathbb{N}$ for which $d(x, y) < \delta$ and $d(T^n(x), T^n(y)) > r$.

The correlated chaos feature refers to the unpredictability behavior of the orbits, in spite of the deterministic rules governing the system. While the definition of chaos in dynamical system is not ubiquitous, here it is presented the most popular one:

Definition 1.3.6 (Devaney's Chaos) A dynamical system (X, T) is said to be chaotic if it satisfies [22, 165]:

- Sensitive dependence on initial conditions
- Topological transitivity, i.e., for any two open sets U and V in the phase space X, there exists a time $n \in \mathbb{N}$ where $T^n(U) \cap V \neq \emptyset$
- Density of periodic orbits, i.e., for any $x \in X$ and $r > 0$, there exists a periodic point $p \in X$ such that $d(x, p) < r$

Apart from the already discussed sensitivity to initial condition, a chaotic system should allow any region of the phase space to be visited by the time evolution of points from any other region (topological transitivity), and it should contain periodic orbits in all the space, so that every point in the phase space is arbitrarily close to one of these. The three conditions above ensure that the system is both unpredictable and complex while retaining some regular periodic behavior.

Chaos or sensitive dependence on initial conditions is usually retrieved by studying long-term behavior, especially on objects called attractors.

Intuitively, an attractor is a set of states (points in phase space) toward which the system evolves over time, representing the long-term behavior of a dynamical system, and is key to understanding its stability and evolution. More formally, one has the following definition:

Definition 1.3.7 If the phase space X is compact according to a given topology, the attractor $\mathcal{A} \subset X$ [22] is a compact set if there exists an open set $U \supset \mathcal{A}$ such that $T(\bar{U}) \subset U$ and $\mathcal{A} = \cap_{n\geq 0} T^n(U)$, where $\bar{U}$ represents the closure of the set U.

From the formal definition, it can be shown that the attractor $\mathcal{A}$ is invariant under the map T, i.e., $T(\mathcal{A}) = \mathcal{A}$; it is attracting for long times $n \to \infty$—it can attract all the points belonging to its *basin of attraction* $BA(\mathcal{A}) = \cup_{n\geq 0} T^{-n}(\mathcal{A})$—and its minimal since $\mathcal{A}$ is the smallest set with properties above.

Types of attractors can be:

- Fixed points, $\mathcal{A} = x_0 \in X$ where x_0 is a stable equilibrium
- Limit cycles, where $\mathcal{A}$ is a closed orbit, attracting nearby trajectories
- Tori, quasiperiodic attractor
- *Strange attractors*, i.e., a fractal set with chaotic behavior where nearby points diverge over time but remain confined to the attractor's structure

Before proceeding with the discussion, it is important to stress that for metric, measure-invariant dynamical system, there exists a notion of ergodicity similar to the stochastic processes setting (Definitions 1.2.5 and 1.2.6). Analogously to the random case, ergodicity refers to the property that, over long periods of time, the time averages of a system's observable quantities are equivalent to their space (ensemble) averages. This means that, for an ergodic system, the trajectory of the system will eventually visit every region of the phase space in a representative way, according to the system's invariant measure.

Definition 1.3.8 (Ergodic Dynamical System) A metrical and measure-preserving dynamical system (X, T, μ) is said ergodic [22] if for any measurable set $A \in \mathcal{F}$, one has

$$T^{-1}(A) = A \implies \mu(A) = 0 \text{ or } \mu(A) = 1.$$

This definition implies that the system cannot be decomposed into smaller invariant sets with positive measure, and almost every point in X will eventually explore the entire phase space in a way that reflects the statistical properties of the whole space.

For the ergodic systems, it holds the following:

Theorem (Birkhoff Ergodic Theorem) For any function on the phase space $f : X \to \mathbb{R}$ (f is called observable) for which $|f|$ has finite integral, for almost every initial point $x_0 \in X$, the time average of f along the trajectory of x_0 converges to the space average, that is, [86]:

$$\lim_{N\to\infty} \frac{1}{N} \sum_{n=0}^{N-1} f(T^n(x_0)) = \int_X f(x)d\mu(x).$$

Thus, the long-term behavior of the system for almost all initial conditions reflects the average behavior over the entire phase space.

With ergodicity, no matter the starting point, the system will eventually behave in a statistically typical manner, meaning time averages for almost all trajectories will match the space average, provided enough time has passed. Again, ergodicity implies that the system does not settle into a particular subset of states but rather explores the whole space.

The previous theorem is the deterministic counterpart of the ergodicity in random processes. Indeed, excluding a set of null measure, any orbit of the system reflects its statistical properties, requiring only one orbit to describe the whole system behavior.

The qualitative study of the long-term dynamical system can be achieved through several quantifiers, each of which reveals specific dynamical systems' properties. Here is a list of the most important.

For a more comprehensive understanding, refer to the following books [22, 86, 165].

1.3.1 Kolmogorov–Sinai Entropy

A tool to quantify the unpredictability and the chaotic nature of a dynamical system in terms of the information needed to predict typical dynamical systems orbits is the Kolmogorov–Sinai entropy. In order to define it, it is necessary to introduce the concept of partition, common refinement of partitions, and information of a state or phase space. The definitions below follow [22].

Definition 1.3.9 (Finite Partition) In a metrical probability space (X, μ), a *partition of* X is defined as a finite collection ζ of sets C_i with the following properties:

- Measurable, i.e., $(C_i) \subset \mathcal{F}$
- Essentially pairwise disjoint, i.e., $\mu(C_i \cap C_j) = 0$ for any $C_i \neq C_j$
- Their union cover X almost everywhere, i.e., $\cup_i C_i = X$ and $\mu(X \setminus \cup_i C_i) = 0$

In substance, a finite partition is a subdivision of the state space into subsets. If two finite partitions ζ and ζ' are defined on X, then it is possible to define a finer new partition, called *common refinement*.

Definition 1.3.10 A common refinement $\zeta \vee \zeta'$ of two finite partitions ζ and ζ' is a finite partition on X composed of all the intersections $C_i \cap C_j$, where $C_i \in \zeta$ and $C_j \in \zeta'$.

Then it is necessary to introduce a measure of the information content in the metric space according to its probability measure.

Definition 1.3.11 (Information Function) Given a metrical space (X, μ), the information function $\mathcal{I}$ is a function $\mathcal{I} : X \to \mathbb{R}$ which has the following properties:

- $\mathcal{I}$ is nonnegative.
- $\mathcal{I}$ is decreasing of the measure of an event $\forall A, B \subset X$ with $\mu(A) \leq \mu(B)$; then $\mathcal{I}(A) \geq \mathcal{I}(B)$.
- $\mathcal{I}(X) = 0$.
- $\mathcal{I}$ is additive for independent sets, i.e., $\mathcal{I}(C \cap D) = \mathcal{I}(C) + \mathcal{I}(D)$ for every $C, D \subset X$ with $C \cap D = 0$.

In this definition, it emerges that higher informational content of observing that event is associated with its lower probability of happening. A suitable choice for $\mathcal{I}(C)$ is $-\log \mu(C)$, for every $C \subset X$.

Remark 1 The previous definitions are still valid if one considers μ a generic measure on X.

In this context, it results natural to quantify how much uncertainty or unpredictability is associated with a given partition if the latter is used to describe the current system's state, in terms of the position occupied by its orbit on the phase space. This can be achieved through the following:

Definition 1.3.12 (Entropy of a Partition) Let ζ be a partition of a probability space (X, μ). The entropy of a finite partition ζ is defined as

$$H(\zeta) = -\sum_{C \in \zeta} \mu(C) \log(\mu(C)). \tag{1.12}$$

The entropy of a finite partition can be seen as the expected value of the information function $\mathcal{I}$ on the elements of the partition.

To capture the role of the dynamics in the orbits predictability according to a fixed partition ζ, one can define a refined finite partition

$$\zeta^n = \zeta \vee T^{-1}(\zeta) \vee \cdots \vee T^{-n+1}(\zeta),$$

where $T^{-k}(\zeta) = \{T^{-k}(C_i) | C_i \in \zeta\}$ is a finite partition. The refined partition ζ^n captures how the states of the system evolve over n time steps, creating a finer grid of possible future outcomes.

Now, the Kolmogorov–Sinai entropy can be formalized.

Definition 1.3.13 (Kolmogorov–Sinai Entropy) Let (X, T, μ) be a dynamical system where the map T is a measure-preserving transformation. The *Kolmogorov–Sinai (KS) entropy* is defined by

$$h_\mu(T) = \sup_{\mathcal{P}} \lim_{n\to\infty} \frac{1}{n} H\left(\bigvee_{i=0}^{n-1} T^{-i}\mathcal{P}\right), \tag{1.13}$$

where the supremum is achieved over all the possible finite partitions on X, according to the probability measure μ.

The limit in (1.13) can be proved to exist thanks to the subadditivity of the entropy H on refined partitions, while the supremum can be theoretically achieved by choosing particular partitions, called *generating* [22].

The KS entropy is the index of the rate of information generation per unit time in the system. It is also the limit of the growth of the partition entropy as the system evolves. Equivalently, KS entropy quantifies the rate at which orbits of the system diverge in state space.

Systems with positive KS entropy show exponential growth of small uncertainties (a hallmark of chaos) even if perfect knowledge of the current state is available. On the other hand, systems with zero KS entropy are either non-chaotic or only exhibit very limited types of complexity (e.g., periodic or quasiperiodic systems).

1.3.2 Lyapunov Exponents

Lyapunov exponents provide important feature of the dynamical system behavior. They measures how quickly a small perturbation δ in the initial condition grows (or shrinks) as the system evolves through iterations of the map T. Their positiveness is usually a signal of sensitive dependence on initial conditions, letting small errors grow quickly, making long-term prediction difficult or impossible, a trait of chaos [22].

To be more precise, Lyapunov exponents' definition relies on assuming that the distance between map iterates of two points x and $x+\delta$ in an open set $U \subset X$, after n steps $\|T^n(x+\delta) - T^n(x)\|$, grows exponentially over time like $\delta \exp(n\lambda)$ for time $n \to \infty$. In this context, one has the following definition:

Definition 1.3.14 Let (X, T, μ) be a metrical dynamical system with T a differentiable map on an open set X. The Lyapunov exponent $\lambda(x, v)$ of the orbit which starts from the initial condition x along the direction $v \in \mathbb{R}^m$ on the tangent space of X is defined as

$$\lambda(x, v) := \limsup_{n\to\infty} \frac{1}{n} \log\|\mathrm{d}T^n(x)v\|, \tag{1.14}$$

where $\mathrm{d}T^n(x)$ denotes the derivative of the T^n at point x.

If T has uniformly bounded first derivatives, then $\lambda(x, v)$ is well defined for every $x \in X$ and every nonzero vector v [22]. Moreover, if $X \subset \mathbb{R}$, then with some algebra (definition of derivative of a composite function and additivity property of the logarithm function) Eq. (1.14) becomes

$$\lambda(x) = \lim_{n\to\infty} \frac{1}{n} \log|dT^n(x)| = \lim_{n\to\infty} \frac{1}{n} \log|dT^n(x)| = \lim_{n\to\infty} \frac{1}{n} \sum_{i=0}^{n-1} \log|dT(x_i)|. \tag{1.15}$$

$x_i = T^i(x)$ are the consecutive orbit points.

On the other hand, when the map is defined on $X \subset \mathbb{R}^m$, the evolution of the system can be linearly approximated by the Jacobian matrix $JT(x)$. In this case, a spectrum of Lyapunov exponents $\lambda_1(x) \geq \lambda_2(x) \geq \cdots \geq \lambda_m(x)$ can be detected, where each exponent describes the exponential growth rate of tangent vectors along orbits.

As stated before, Lyapunov exponents are index of stability and chaos of the system. In particular, if Lyapunov exponents are strictly positive, the system results chaotic (at least in the manifold of the corresponding Lyapunov exponent); if $\lambda_i(x) = 0$, then the system is neutrally stable or may exhibit periodic or quasiperiodic behavior in the corresponding tangent direction, and in the tangent directions where $\lambda < 0$, the system shows stability, meaning nearby trajectories converge, and perturbations decay over time.

Lyapunov exponents are also related to the aforementioned KS entropy. Indeed, under some assumptions, it can be proved that KS entropy is the sum of the positive Lyapunov exponents (Pesin's identity [110]).

Other dynamical systems' quantifiers are presented below.

1.3.3 Fractal Dimension

A fractal dimension is a measure used to describe fractal patterns or sets (i.e., geometrical objects, each part of which has the same statistical character as the whole) by quantifying their complexity as the ratio of detail change to scale change. It is used to characterize intricate geometric shapes where fine details are more significant than the overall structure [109]. Unlike classical dimensions (e.g., 1D, 2D, and 3D), the fractal dimension can take non-integer values, reflecting how completely a fractal appears to fill space as it scales.

The fractal dimension is usually employed to study complex and chaotic dynamical systems, which can generate orbits resembling fractals in the phase space. Indeed, it can be a quantitative measure of the system's complexity, being higher fractal dimensions associated with more intricate dynamics, while lower dimensions suggest more predictable, smoother behavior. Moreover, in chaotic systems, the fractal dimension denotes how complex or "dense" are the strange attractors, i.e., geometric sets with a fractal structure on which the dynamics converge for long times.

There exist different notions of fractal dimensions. Here the main ones are listed.

1.3.3.1 Hausdorff Dimension

The Hausdorff dimension is one of the most rigorous and widely used definitions of fractal dimension. It generalizes the concept of dimension from integer values to non-integer values for fractals [151].

Definition 1.3.15 (Hausdorff Dimension) For any set S in a metric space X, the Hausdorff dimension D_H is defined as

$$D_H = \inf\left\{d \geq 0 : \lim_{r\to 0}\left(\inf\sum_i (\text{diameter of covering } U_i)^d\right) = 0\right\}, \tag{1.16}$$

where U_i is a collection of subsets that cover S, and the diameter of a set is the largest distance between two points in the set.

Again, D_H can take non-integer values, reflecting the scale-invariant nature of fractals. While the Hausdorff dimension is often used in theoretical studies of fractal sets and strange attractors in dynamical systems, alternative fractal dimensions are frequently employed due to their computational simplicity and practical applicability.

1.3.3.2 Box-Counting Dimension

The box-counting dimension D_B [58] is one of the simplest and most practical methods to calculate fractal dimension. It approximates the dimension by covering the object with a grid of boxes and counting how the number of boxes needed to cover the object changes with the size of the boxes.

Definition 1.3.16 (Box-Counting Dimension) Let $S \subset X$. If $N(r)$ denotes the number of boxes of side length r required to recover the set S, then the box-counting dimension is defined as

$$D_B = \lim_{r\to 0}\frac{\log N(r)}{\log(1/r)} \tag{1.17}$$

To practically evaluate the box-counting dimension D_B, one overlays a grid on the fractal object, then counts the number of boxes that contain any part of the object; then one repeats the previous two steps by decreasing the size of the boxes and finally one gets D_B computing the slope between $\log N(r)$ versus $\log(1/r)$.

1.3.3.3 Correlation Integral and Correlation Dimension

The correlation integral $C(r)$ is a measure used in the analysis of the spatial distribution of points in a dataset, typically to estimate the fractal dimension, especially the correlation dimension of a chaotic attractor. It quantifies how many pairs of points are close to each other within a given distance r.

If X is provided of the distance d, the correlation integral can be formally described as

$$C(r) = \frac{2}{N(N-1)} \sum_{i=1}^{N-1} \sum_{j=i+1}^{N} H(r - d(x_i, x_j)), \tag{1.18}$$

where N represents the total number of samples of the orbit, x_i and x_j are two orbit points, and $H(x) = \{1, \text{if } x \geq 0\ 0, \text{ if } x < 0\}$ is the Heaviside step function.

The correlation integral is the main ingredient for defining the correlation dimension [54], which is a statistical measure used to estimate the fractal dimension based on how pairs of points in the dataset relate to each other. It is useful for studying the fractal properties of time series data and strange attractors in chaotic dynamical systems.

Definition 1.3.17 Given an orbit of a dynamical system (the same applies to time series), the correlation dimension D_C is given by

$$D_C = \lim_{r \to 0} \frac{\log C(r)}{\log r}, \tag{1.19}$$

where $C(r)$ is the correlation integral.

Hence, for small r, the following scaling relationship holds:

$$C(r) \propto r^{D_C}. \tag{1.20}$$

1.3.4 *Approximate Entropy*

Approximate Entropy (ApEn) is a statistic that quantifies the unpredictability of fluctuations in a time series by assessing the likelihood that patterns that are close to each other will remain close in the future. In some sense, it can be seen as an approximation of the KS entropy [36] for infinite orbits time series. It was introduced by Pincus [132] to measure system complexity and regularity.

Given a finite monodimensional orbit $\{x_1, x_2, \ldots, x_N\}$ of length N (the same holds for any time series), ApEn is computed by comparing sequences of data points to assess the similarity between them, using a predefined tolerance r. The following steps are performed for the ApEn computation.

- For a given embedding dimension m, form vectors $\mathbf{u}(i)$ consisting of consecutive points in the time series:

$$\mathbf{u}(i) = [x_i, x_{i+1}, \ldots, x_{i+m-1}]$$

where $i = 1, 2, \ldots, N - m + 1$. This procedure is called *phase space reconstruction*, and its foundation relies on Takens' embedding theorem [122] which gives the conditions under which a dynamical system can be reconstructed from a sequence of observations of the state of that system. While the reconstruction is a diffeomorphism that preserves continuity and differential behavior of a dynamical system, it does not preserve the geometric shape of structures in phase space.
- Define the distance between two vectors $\mathbf{u}(i)$ and $\mathbf{u}(j)$ as the maximum absolute difference between their corresponding components:

$$d(\mathbf{u}(i), \mathbf{u}(j)) = \max_{k=1,\ldots,m} |x_{i+k-1} - x_{j+k-1}|. \tag{1.21}$$

- For each i, count the number of j's (for $j = 1, 2, \ldots, N - m + 1$) such that the distance between $\mathbf{u}(i)$ and $\mathbf{u}(j)$ is less than or equal to the tolerance r. This count is denoted by

$$C_i^m(r) = \frac{\text{number of } j \text{ such that } d(\mathbf{u}(i), \mathbf{u}(j)) \leq r}{N - m + 1}. \tag{1.22}$$

- The natural logarithm of the average value of $C_i^m(r)$ over all i is computed as

$$\Phi^m(r) = \frac{1}{N - m + 1} \sum_{i=1}^{N-m+1} \ln C_i^m(r).$$

- Approximate entropy is then defined as the difference between the values of $\Phi^m(r)$ for embedding dimensions m and $m + 1$:

$$\text{ApEn}(m, r, N) = \Phi^m(r) - \Phi^{m+1}(r).$$

Note that different distances can be defined on reconstructed vectors space (e.g., Euclidean norm, or p-distances induced by p-norm).

The smaller the value of ApEn, the more predictable or regular the time series. Larger values indicate more complexity or irregularity.

1.3.5 Sample Entropy

Sample Entropy (SampEn) is a modification of Approximate Entropy (ApEn) designed to measure the regularity and complexity of a time series [138]. Unlike ApEn, SampEn does not count self-matches, which reduces bias and makes it more robust for smaller datasets. It quantifies the likelihood that similar patterns of observations will remain similar in the future, with a smaller SampEn indicating more regularity and a larger SampEn suggesting more complexity.

Given a finite, monodimensional orbit $\{x_1, x_2, \ldots, x_N\}$ of length N, SampEn is defined as the negative natural logarithm of the conditional probability that two sequences close to each other for m points remain close for $m + 1$ points, within a given tolerance r.

The steps for reconstructing SampEn are similar to the ones for the ApEn. Indeed, after the embedded procedure for reconstructing the m-dimensional space, thus forming the embedded vectors $\mathbf{u}(i)$, with $i = 1, 2, \ldots, N - m + 1$, and after defining a proper distance d between vectors, one has to:

- Count similar pairs for length m, by counting, for each i, the number of vectors $\mathbf{u}(j)$ (for $j \neq i$) such that

$$d(\mathbf{u}(i), \mathbf{u}(j)) \leq r.$$

  Denote this count by $B_i^m(r)$. Then, the average count of similar vectors for all i is

$$B^m(r) = \frac{1}{N - m} \sum_{i=1}^{N-m} \frac{B_i^m(r)}{N - m - 1}.$$

- Repeat the previous process for $m + 1$, forming $B^{m+1}(r)$.
- Define the Sample Entropy as the negative natural logarithm of the ratio between $B^{m+1}(r)$ and $B^m(r)$:

$$\text{SampEn}(m, r, N) = -\ln\left(\frac{B^{m+1}(r)}{B^m(r)}\right).$$

Again, a smaller value of SampEn indicates a more regular, predictable time series, while a larger value suggests greater complexity or irregularity. Because SampEn does not include self-matches (i.e., $i = j$), it provides a less biased estimate of complexity compared to Approximate Entropy (ApEn), especially in shorter time series.

Chapter 2
Formal Definition of Noise and Noise Estimation Techniques

Abstract In this chapter, the concept of noise is introduced both as a stochastic process—defined as a sequence of random variables—and as a perturbative element within a dynamical system. A classification of noise is then provided based on its frequency content, commonly referred to by its "color", and a distinction is made between output noise and dynamical noise. The latter, being embedded within the system's evolution, fundamentally alters its dynamics and can be interpreted as information intrinsic to the system that cannot be inferred solely from its past behavior.

The chapter focuses primarily on dynamical noise, presenting a survey of existing methods for its estimation within dynamical systems. Main approaches including nonlinear time series analysis, stochastic differential equations, probabilistic-Bayesian frameworks, and state-space modeling, including the well-known Kalman filter, are briefly reviewed.

Keywords Noise · Additive noise · Output noise · Dynamical noise · Dynamical noise estimation techniques

2.1 Noise as a Pure Random Process

In this section, the main types of noise are defined and characterized based on their statistical properties. It is useful to describe the characteristics of noise in terms of its power spectral density (PSD), which quantifies how the power—or variance—of a signal or process is distributed across different frequency components. The PSD provides valuable insight into the frequency structure of the noise and is frequently used in the classification and analysis of stochastic processes.

Definition 2.1.1 (Power Spectral Density) Let $Y(t)$ be a continuous-time wide-sense stationary stochastic process with mean $\eta = \mathbb{E}[Y_n]$ for all admissible t and autocovariance function $\gamma_Y(\tau)$, being τ the lag between two time points. The power spectral density (PSD) $S_Y(f)$ of the process [164] is defined as the Fourier transform

A. Scarciglia et al., *Physiological Noise and Informative Randomness in Neural and Cardiovascular Systems*, https://doi.org/10.1007/978-3-031-95914-1_2

of its autocovariance function

$$S_Y(f) = \int_{-\infty}^{\infty} \gamma_Y(\tau)e^{-if\tau}d\tau, \tag{2.1}$$

where f is the angular frequency and i is the imaginary unit.

It can be shown that $S_Y(f) \geq 0$, that $S_Y(f) = S_Y(-f)$ for real-valued process Y_n, and that the total power of the process can be obtained by integrating the PSD over all frequencies

$$\text{Var}(Y_t) = \gamma_X(0) = \int_{-\infty}^{\infty} S_Y(f)df$$

by using Parseval's theorem.

The PSD can be defined for discrete-time stochastic processes as well, replacing the integral and the Fourier transform of Eq. 2.1 with the sum over the integer lags τ and the discrete-time Fourier transform.

In the following sections a list of the main noise processes differentiated by "color" is presented, which refers to the distinct frequency characteristics, often named after colors due to analogies with light spectra. The "color" of noise describes the distribution of its power across different frequencies, which influences the signal's properties and behavior.

The characteristics of noise are reported within discrete-time stochastic processes framework, whereas it is possible to extend the definition of noises in the continuous-time domain. This choice is motivated by the fact that signals under study are usually discrete-time due to the sampling procedure. To do so and in order to improve clarity of the discussion, the following notation is introduced: An infinite sequence is denoted as $\{a_n\}_{n\geq 0}$, where the elements a_n may represent either discrete-time stochastic processes or deterministic sequences, such as the orbit of a dynamical system. For simplicity, this notation is abbreviated as $\{a_n\}_n$, unless stated otherwise. If a sequence contains a finite number of samples, N, then it is represented as $\{a_n\}_{n=0}^N$.

2.1.1 White Noise

White noise [12, 91] is a type of stochastic process or a signal that is characterized by having a constant power spectral density over all frequencies, meaning that it contains equal intensity at different frequencies. It is widely used in various fields, including signal processing, statistics, and physics [91]. Formally, one has the following definition:

Definition 2.1.2 (White Noise) A discrete-time stochastic process $\{\varepsilon_n\}_n$ is called *white noise* if the process has $\mathbb{E}[\varepsilon_n] = 0$, $\text{Var}(\varepsilon_n) = \sigma^2$ for all n, and $\text{Cov}(n_1, n_2) = 0$ for $n_1 \neq n_2$.

It follows that the PSD of white noise is uniform across all frequencies $S_\varepsilon = \sigma^2/2\pi$.

In practical scenarios, it is often convenient to assume that the distribution of white noise samples is Gaussian. This assumption is justified because, in many real-world cases, noise results from the combination of numerous small, independent random factors (such as sensor inaccuracies or environmental influences). According to the Central Limit Theorem, the sum of these factors tends to follow a normal distribution, making the Gaussian assumption a reasonable and widely applicable choice. Hence, one has the following definition:

Definition 2.1.3 (White Gaussian Noise) A random process $\{\varepsilon_n\}_n$ is said white Gaussian noise (WGN) if it is a white noise where all the composing random variables follow a Gaussian distribution $\varepsilon_n \sim \mathcal{N}(0, \sigma^2)$.

It can be shown that WGN processes are also IID.

WGN processes are fundamental models used in communication systems, to represent thermal noise affecting signals, in signal processing, to simulate random noise or disturbances and in control systems as an idealized general noise model.

2.1.2 Pink Noise Processes

Pink noise processes [168], also known as $1/f$ processes, describe stochastic or random processes where the PSD decreases with increasing frequency [168]. Specifically, pink noise exhibits a frequency distribution where the power decreases inversely with frequency, making it distinct from white noise, which has equal power across all frequencies.

Definition 2.1.4 (Pink Noise) A pink noise process can be modeled as a random process $\{\varepsilon_n\}_n$ where the PSD follows the $\frac{1}{f}$ law. Formally, the PSD $S_\varepsilon(f)$ scales like $1/f$, while the autocovariance function $\gamma_\varepsilon(\tau) \propto \frac{1}{\tau^\beta}$, $0 < \beta < 1$.

Pink noise processes exhibit long-term dependencies and correlations. In contrast to white noise, where each sample is independent, pink noise has correlated samples, meaning that past values influence future values. This results in smoother, more gradual fluctuations over time compared to white noise, which appears more random and spiky. Due to the long-range correlations, the autocovariance function of pink noise decays more slowly compared to white noise. In fact, the autocovariance function does not drop off sharply but decays following a power law, reflecting the persistent correlations in the process.

Pink noise processes appear frequently in natural and man-made systems. Some examples include the following: fluctuations in human heartbeats often follow $1/f$ dynamics [168]; temperature variations and long-term climate data exhibit pink noise characteristics; financial data, such as stock price fluctuations, sometimes show behavior consistent with pink noise; and many naturally occurring sounds, including music and human speech, follow a $1/f$ frequency distribution.

2.1.3 Brownian Noise processes

Discrete-time Brownian noise processes [106], also known as discrete-time random walk processes, are stochastic processes that represent the discrete-time counterpart of continuous Brownian motion [46]. These processes are characterized by their cumulative nature, where each value is the sum of independent, normally distributed random variables, making them an example of a random walk.

Definition 2.1.5 (Brownian Noise) A discrete-time stochastic process $\{B_n\}_n$, with $B_0 = 0$, is said Brownian noise if at the time n, the process increments according to a normal distribution:

$$B_n = B_{n-1} + \varepsilon_n,$$

where $\varepsilon_n \sim \mathcal{N}(0, \sigma^2)$.

It follows that Brownian noise samples have $\mathbb{E}[B_n] = 0$, $\gamma_B(n_1, n_2) = \sigma^2 \min(n_1, n_2)$ (i.e., the values of the process at different times are highly correlated), and $S_B(f) \propto \frac{1}{f^2}$, reflecting the dominance of slow fluctuations in the process. This noise process is widely used to model phenomena in finance (stock prices) and physics (random motion of particles suspended in a fluid).

2.1.4 Other Colored Noises

In addition to the important noisy processes previously described, there are other types of colored noise processes, such as blue [180], violet, black, and gray noise, which are, respectively, characterized by a PSD $S(f) \propto f$, $S(f) \propto f^2$, $S(f) \propto 1/f^{\alpha}$ with $\alpha > 2$ and $S(f)$ tailored for human hearing [52].

2.1.5 Other Noise Processes

Besides noise processes classification in terms of color, or spectral properties, it is also possible to distinguish these random processes by considering the distribution of their samples. In particular, one has Gaussian, Poisson, and Cauchy processes whose joint PDF follows, respectively, jointly Gaussian [178], Poisson, and Cauchy distributions.

The former, as already specified, is assumed in many systems because of its mathematical tractability and the Central Limit Theorem (small, independent noise sources aggregate into Gaussian noise), and it models thermal noise in electronic circuits and simulates sensor noise in digital image.

The Poisson noise [163] occurs in systems where events are random and discrete over time, and the number of events in each time interval follows a Poisson distribution. It is mainly used in optical systems (e.g., in astronomy, microscopy, or medical imaging) where the number of detected photons over time is modeled by Poisson random variable, and it models shot noise in semiconductor devices.

The Cauchy noise process [182], instead, is a type of heavy-tailed noise where samples are drawn from a Cauchy distribution, characterized by frequent large outliers due to its undefined mean and variance. It is mainly used in systems prone to large outliers and for modeling impulsive or heavy-tailed noise.

Other important types of noise that are treated here are:

- Transient noise [35] which refers to short-duration, high-energy noise events that temporarily disrupt a signal or a system. These disturbances typically occur in nonstationary processes, lasting for brief periods before the system returns to its normal state.
- Burst noise [39], which consists of random, discrete bursts of noise that occur over time, often with varying amplitude and duration. It is observed primarily in semiconductor devices.
- Phase noise [49], a type of noise that affects the phase of a signal, especially in oscillators or signal sources. It causes small, rapid fluctuations in the phase of a waveform, which degrades the signal's frequency stability.
- Quantization noise [55], arising when a continuous analog signal is converted into a discrete digital signal. The difference between the actual analog value and the closest representable digital value introduces noise.

2.2 Output and Dynamical Noise

In Sect. 2.1 it has been provided a pure statistical definition of noise, which mainly relies on the correlations among its samples and on the consequent power spectral densities properties.

As already discussed, independently on the fact that noise is considered a disturbance or an intrinsic component of a system, it is undoubted that its presence modifies the observed trajectory of the phenomenon under study.

In this context, two more definitions of noise which rely on its effects on the dynamics are provided. The following definitions are taken from [84].

Definition 2.2.1 (Output Noise) Let $\{x_n\}_{n\geq 0}$ be an orbit of a dynamical system (X, T, μ) starting from an initial state x_0, whose elements satisfy $x_{n+1} = T(x_n)$ for all n. A random process $\{\varepsilon_n\}_{n\geq 0}$ is said to be *output noise* for an orbit $\{x_n\}_{n\geq 0}$ if it produces an orbit of the following form:

$$y_n = x_n + \varepsilon_n$$

for all n.

Definition 2.2.2 (Dynamical Noise) Let $\{x_n\}_{n\geq 0}$ be an orbit or trajectory of a dynamical system (X, T, μ) starting from an initial state x_0. A random process $\{\varepsilon_n\}_{n\geq 0}$ is said to be *dynamical noise* for a trajectory of $\{x_n\}_n$ if it produces a trajectory $\{y_n\}_n$ of the following recursive form:

$$y_n = T(y_{n-1}) + \varepsilon_n \tag{2.2}$$

for all n, with $y_0 = x_0$.

The former can occur during the process of measuring a system's state, and it can be due to sensor inaccuracies, environmental interference, or random fluctuations in the measurement process, affecting the observed measurements and introducing errors in the estimated state. The system's dynamics are not eliminated but rather obscured by random fluctuations, which can be easily removed in many cases.

The dynamical noise, instead, arises from inherent uncertainties or fluctuations *within* the system itself and it may significantly affect a deterministic system's state trajectory. It becomes challenging to gain a clear understanding of the actual phenomenon being studied.

Because of its stochastic nature, for its non-time-punctual predictability, quantities such as mean and median values, as well as variance and others, are often exploited for its characterization, which might be only obtained by observations over time.

Given the significant role of noise as an inherent information source within physiological processes, revealing dynamics beyond historical predictability, we will model the system with dynamical noise.

Remaining as general as possible, it is assumed that the dynamical noise is modeled as a sequence of IID random variables $\{\varepsilon_n\}_n$, where each $\varepsilon_n \sim \mathcal{N}(0, \sigma^2)$ (i.e., it is white Gaussian noise with zero mean and fixed standard deviation) that is also involved into the dynamics according to Eq. 2.2. This generality follows from the already discussed properties of the white noise (see Sect. 2.1.1), which results from the summation of small, independent stochastic perturbations in virtue of the Central Limit Theorem.

A schematic representation of these two noises and their interactions with the dynamics is represented in Fig. 2.1, while exemplary neural time series corrupted by output and dynamical noises are illustrated in Fig. 2.2.

Efforts have been made to remove the two kinds of noise for retrieving the original/deterministic signal underlying the system dynamics. While measurement noise can be relatively easily removed by studying sensor inaccuracies or the environmental interference, dynamical noise reduction is not a trivial task [154], especially when the system's dynamics is unknown or when the noise levels are high to drastically change the underlying dynamics.

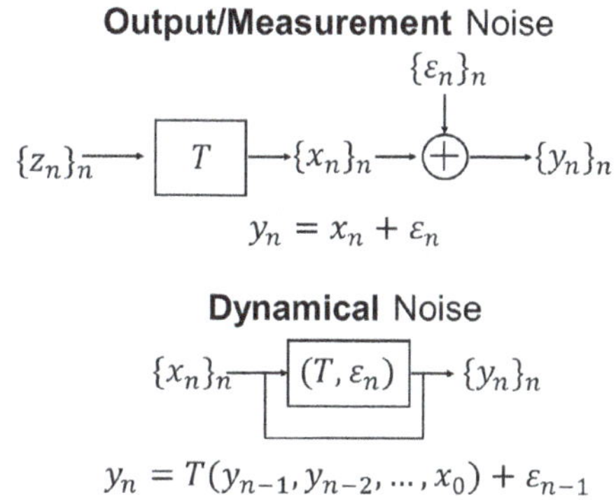

Fig. 2.1 Schematic representation of output and dynamical noises. Lock diagram of noise interactions with the underlying dynamics in the case of output noise (upper panel) and dynamical noise (lower panel). In the former, the noisy process $\{\varepsilon_n\}_n$ is added to an unperturbed orbit; in the latter, each noise sample ε_n is integrated into the dynamics, significantly altering the output

Fig. 2.2 Neural signals affected by dynamical and output noise. Exemplary neural dynamics (tonic spiking simulated using the Izhikevich model [73]) in the case of no noise (**a**) and with additive (**b**) and dynamical (**c**) noise whose samples ε_n are pairwise uncorrelated (white noise) with Gaussian distribution of mean 0 and $\sigma = 10$[a.u.]

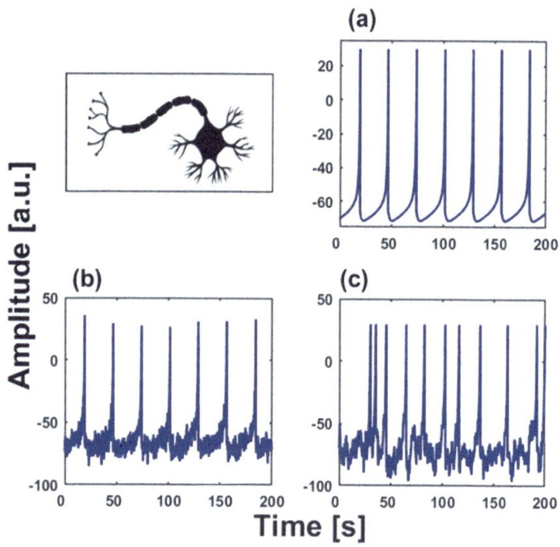

2.3 Dynamical Noise Estimation Methodologies

In this section, the methods used to estimate dynamical noise from an observable time series are listed. Due to the ubiquitous presence of noise in various scientific fields, numerous studies have been conducted on this topic. Using the *Scopus* database, the following keywords have been selected:

- ("dynamical noise" OR "dynamical noise levels") AND ("time series" OR "estimation" OR "quantification" OR "detection" OR "determination")

- ("chaotic time series" OR "chaotic dynamical systems") AND ("noise level" OR "noise" OR "dynamical noise" OR "noise level estimation") AND ("estimating" OR "estimation" OR "determination" OR "quantification")
- ("process noise covariance estimation" OR "process noise estimation")

Subsequently, only papers published in journals were selected, excluding articles in which dynamical noise was not directly estimated or where other methods were shown to be robust against this type of noise.

2.3.1 Nonlinear Estimation Methods

The methods presented in this section exploit nonlinear and chaotic dynamical system theory. The main tools are:

- The correlation integral and correlation dimension, previously described in Sect. 1.3.3.3
- Lyapunov exponents, previously analyzed (Sect. 1.3.2)
- The correlation entropy [53, 124], which measures the average information content per data point in the reconstructed m-dimensional phase space X_{emb} of the system. It is closely related to the correlation dimension and can be calculated from the correlation integral. It is formally computed as

$$K_2 = \lim_{r\to 0}\lim_{N\to\infty}\lim_{m\to\infty}\frac{\ln C_m(r)}{m},$$

  where N is the total number of points, and $C_m(\varepsilon)$ is the correlation sum (or integral) of the reconstructed space of embedding dimension m.
- Recurrence plots, a powerful tool that visualizes the times at which the state of a dynamical system revisits the same or a similar state, offering insights into the system's behavior over time [41]. Formally, after embedding the series in an m-dimensional space X_{emb}, recurrence plots are represented by a square matrix R where the element $R(i, j)$ is 1 if the states x_i and x_j are within a specified threshold distance r, and 0 otherwise, as defined by

$$R(i,j) = \begin{cases} 1, & \text{if } ||x_i - x_j|| \leq r \\ 0, & \text{otherwise.} \end{cases}$$

Unless specified otherwise, the following methods assume that the observable time series x_n is generated by a perturbed dynamical system

$$x_n = T(x_{n-1}) + \varepsilon_n,$$

where the map T defines the system model, and ε_n represents realizations of a stochastic Gaussian white noise variable $\mathcal{N}(0, \sigma)$. The term "noise estimation" refers to the quantification of the σ value.

In [181], the authors present a noise estimation technique based on the functional dependence between the coarse-grained entropy $K_2(\varepsilon)$ and the threshold ε, as well as the correlation dimension, which incorporates the noise standard deviation within the time series. A similar method is proposed in [90], where the authors suggest a correction to the embedding dimension estimate in noisy regimes.

The work in [169] also offers an estimation of the dynamical noise in the correlation integral and the K_2 entropy, utilizing geometric considerations on how noise distributes in the reconstructed phase space along the attractor—a subset of the phase space where the dynamics lie after a transient period.

Other noise estimation methods specifically designed for estimating measurement or output noise in chaotic time series, and also applicable to infer dynamical noise levels, are also presented here. The main noise identification procedure is found in [154], where the noise level in time series is determined by linking the scaling properties of the correlation integral to a Gaussian noise functional involving the first two moments of the Gaussian noise distribution. This method was further refined in [99, 194] to detect higher levels of noise by computing a similar estimator of the correlation integrals and sums using the L^∞-norm, instead of the Euclidean L^2-norm in the phase space, in order to reduce the effects of the attractor's curvature. A further refinement of this noise level estimation is provided in [75] through the correlation dimension and noise level estimations using the linear maximum likelihood procedure or the linear least-squares method [76], which reduces large estimation errors. In [29], noise estimation is achieved by representing Schreiber's Gaussian functional in terms of confluent hypergeometric functions of the first kind and by developing a linear least-squares algorithm based on a truncated version of the functional, which reduces computational costs.

Another approach to noise estimation in time series involves measuring the noise drift of the largest Lyapunov exponent [195, 196], a nonlinear index of orbit divergence in a dynamical system [36].

2.3.2 *Probabilistic Approaches*

Other methods for noise identification are purely based on probability theory. In particular, they rely on Bayes' theorem, which retrieves the noise level σ as the most likely value given the model and data distribution. These methods also employ Monte Carlo simulation techniques, which use random numbers to estimate numerical results (such as statistical mechanics quantities), and probability density function estimations.

The Bayesian probabilistic approach is used to detect and distinguish output or measurement noise from dynamical noise, provided that there is a reliable fit or approximation of the underlying deterministic model [61]. This method has also been applied for time series forecasting in regimes dominated by dynamical noise [64]. In [166], the power of dynamical noise was identified and distinguished from measurement noise in Hénon and Ikeda maps, revealing that the probability density function of time series samples corrupted by dynamical noise resembles a Cauchy distribution. This method further exploits the scaling properties of correlation entropy, which primarily depends on the distribution of additive noise contaminating a given time series and appears to be independent of the dynamical noise distribution. Similarly, the Markov Chain Monte Carlo Bayesian approach described in [82] enables dynamical denoising by providing an iterative estimation of the dynamical noise within the series.

2.3.3 *Stochastic Differential Equation (SDE) Formalism*

Dynamical noise levels can also be formalized and quantified within the framework of continuous-time stochastic differential equations. In this context, a general form of a noise-drifted process can be expressed using Langevin diffusion processes [96]:

$$dx(t) = f(x(t))dt + \sigma dB(t)$$

where $x(t)$ represents the multidimensional state of the system at time t, $f(x(t))$ is the deterministic force acting on the system, σ is a constant that determines the strength of the random fluctuations (noise level), and $dB(t)$ is a Wiener process, which is a continuous-time random walk with independent increments. The first term on the right-hand side of the equation represents the deterministic evolution of the system, while the second term accounts for random fluctuations.

In [158], a model-free noise quantification method for both output and dynamical noise was proposed through the reconstruction of Langevin processes. This approach utilizes the Kramers–Moyal coefficients [119] and conditional moments [20], applied to sufficiently long time series. Within the Langevin framework, measurement noise interferes with dynamical noise detection by introducing estimation drifts, generating offsets in the process. A similar effort to distinguish output noise from dynamical noise was undertaken through statistical tests in [19], specifically for the Van der Pol oscillator.

In contrast, the authors of [144] provided an estimation of dynamical noise levels, scaled from 0 to 1, within the Wiener process framework by employing the time series dimension (TSD) [63], a technique used to measure the fractal dimension of attractor segments in phase space. Their method relies on an analytical relationship between the TSD and the levels of dynamical noise.

2.3.4 Kalman-Like Filters

Dynamical noise can be referred to as process noise. In this framework, dynamical systems are typically represented as

$$\begin{cases} x_{k+1} &= f(x_k, t_k) + w_k \\ y_k &= g(x_k, t_k) + v_k, \end{cases} \tag{2.3}$$

where x_k are the n-dimensional state vectors and y_k are m-dimensional observation vectors. The functions f and g define, respectively, the underlying model and the observation function, which represents what can actually be measured. The terms w_k and v_k are samples of noisy distributions: The former arises from process (dynamical) noise since it interacts with the system's dynamics, while the latter is associated with output/measurement noise. Both noise samples are assumed to be drawn from white noise distributions, with covariances Q_k and R_k, respectively.

The primary goal is typically to estimate the states x_k given the measurements y_k. This procedure also acts as a denoising technique, as it mitigates the effects of noise on the observed data.

In [80], Kalman proposed a filter that allows one to retrieve the states x_k if the functions f and g are linear, provided that the process and measurement noise covariances are properly tuned. The filter operates in real time, recursively, and works through a two-phase process: a prediction phase and an update phase. In the prediction phase, the Kalman filter generates estimates of the current state variables along with their uncertainties. In the update phase, once the next measurement is observed, these estimates are refined using a weighted average.

The Kalman filter has since been extended to handle nonlinear systems (Extended and Ensemble Kalman filters) [11, 81] while still requiring knowledge of the system's equations and noise levels. Some variants of the Kalman filter (known as unscented filters) have been applied when no prior knowledge of the model is available [16].

Nevertheless, the challenge of tuning the noise covariance matrices Q_k and R_k, which are critical to state estimation, remains. A common approach is to a priori tune these noise matrices and then apply the Kalman filter to retrieve the underlying dynamics. However, these procedures may affect the accuracy of state estimation, potentially causing the filter to diverge when large errors exist in the process noise covariance matrix.

Many techniques for estimating both the state and process noise are variations of adaptive Kalman filters, which can adjust their parameters to account for changes in system dynamics or measurement noise characteristics. This adaptability is especially useful for systems that are nonstationary or have uncertainties that evolve over time.

For instance, the process noise covariance (or dynamical noise level) has been estimated in [162] through adaptive filters, where dynamical constraints are also imposed on the noise; in [67], an adaptive unscented Kalman filter was used;

in [44], an adaptive filter based on reinforcement learning was proposed; and in [113], a filter for noise estimation was applied online, achieved by comparing expected and observed event rates using the golden section search optimization algorithm. A moving-window version of the extended Kalman filter has also been proposed [94], which has been used for modeling process noise in structural damage detection. Other noise estimation methods include iterative low-rank matrix approximation [101], Bayesian probabilistic theory [16], stochastic approximation [14], and adaptive cubature information filters [78].

Additionally, a further extension of the Kalman filter, known as the Kalman–Takens filter, has been proposed to address the lack of a model and the need for a priori noise tuning [59]. This filter reconstructs the system in a higher dimensional space based on Taken's embedding theory [146], provided there is a sufficient number of time series samples. This approach bridges nonlinear analysis tools and classical signal processing filtering techniques. Moreover, it eliminates the necessity for the process and measurement noise matrices Q_k and R_k to recover the system's states. Instead, noise covariances can be estimated independently through adaptive techniques that iteratively adjust the noise matrices from an initial random guess. This method has also proven effective for dynamical noise reduction when compared with classical Kalman filters.

In [98], a statistical estimator for linearly autocorrelated noise is proposed. Impulsive noise, defined as instantaneous disturbances or artifacts, is detected in [70] using nonlinear prediction or adaptive Kalman filters [173].

2.3.5 *Considerations on the Dynamical Noise Estimation Techniques*

Previous sections reviewed existing methods for estimating dynamical noise, drawing upon nonlinear analysis, Bayesian theory, probability theory, and dynamical systems modeling (e.g., stochastic differential equations and state-space models like Kalman filters). While these methods offer valuable insights, they often exhibit limitations, such as strong dependence on system dynamics, restricted estimation ranges, or a primary focus on output noise estimation. Some studies, in addition, primarily examine the dependence on noise of some quantifiers (e.g., Lyapunov exponents or correlation dimension). Furthermore, the philosophy underpinning the previously described methods is to consider noise as a source of disturbance or an exogenous component of the signal that must be identified and, if possible, removed from the system. In light of these limitations, the following chapters will introduce novel methods specifically designed for estimating dynamical noise, particularly physiological noise arising from biological recordings: The identification of dynamic/physiological noise in biological systems will unlock new insights into the variability of physiological system dynamics, revealing its origins and providing a deeper understanding.

Chapter 3
Physiological Noise in Neural and Cardiovascular Systems

Abstract In this chapter, the concept of physiological noise is formally introduced. Beginning with the assumption that a physiological system can be represented as a dynamical system governed by an unknown analytical map, physiological noise is modeled as dynamical noise acting upon this system—i.e., as a stochastic component intrinsically embedded in its dynamics.

The chapter then provides an anatomical overview of the physiological systems examined in this book, with a focus on neuro-autonomic cardiac control and neural systems across multiple scales, including individual neurons, the brain, and the cerebral cortex. For each of these, the signals used for investigation are discussed, along with the primary sources of physiological noise inherent to each system.

Keywords Physiological noise · Cardiovascular systems · Heart rate variability · HRV · Neuronal activity · Brain dynamics

3.1 Physiological Noise Definition

Consider a generic physiological system as a discrete metrical and measure-preserving dynamical system (X, T, μ), where X is a compact subset of $\mathbb{R}$, and T is any differentiable map function with bounded derivative, which preserves the probability measure μ. A typical noise-free output of this system is represented as $w_n = T(w_{n-1}, w_{n-2}, \ldots, w_0)$, where $w_i \in X$ for all positive integers i.

Definition 3.1.1 (Physiological Noise) Under the previous setting, the *Physiological Noise* is defined as a sequence of IID random variables $\{\varepsilon_n\}_n$, which constitutes the *dynamical noise*, whose samples ε_n, in fact, modify physiological dynamics at each step according to the equation:

$$x_n = T(x_{n-1}; x_0) + \varepsilon_n, \tag{3.1}$$

A. Scarciglia et al., *Physiological Noise and Informative Randomness in Neural and Cardiovascular Systems*, https://doi.org/10.1007/978-3-031-95914-1_3

where $x_i \in X$ for all positive integers i, where the map T considers the past history of the system, from the sample x_{n-1} up to the initial condition $x_0 = w_0$, here denoted by $(x_{n-1}; x_0)$.

A finite physiological time series contaminated by physiological noise $\{\varepsilon_n\}_n$ can be denoted as $\{x_n(\varepsilon)\}_{n=0}^{N}$—this notation can be extended to any time series. From the best of our knowledge, without any specific information or assumption on T, it is not possible to estimate the noise ε_n formally.

In the following sections, the neurocardiovascular and neural systems will be anatomically described at different scales, from neurons to the brain, with a focus on the etiology of noise within these systems.

3.2 Physiological Noise in the Cardiovascular System

3.2.1 Overview of the Cardiovascular System

The cardiovascular system is part of the circulatory system, and it carries blood through the whole human body, containing nutrients, immune functional cells, and waste that maintain homeostasis and basic functions of cells and organs [112, 123]. The system is formed by the heart, an anatomical pump, and intricate conducts, namely veins, arteries, and capillaries. The pumping action of the heart takes places in the two ventricles (one per each left side and right side), lower chambers where the blood enters after the passage through the action of two other upper chambers, the atria. The chambers are separated by a wall, and the passage of the blood through them is controlled by valves. By actively contracting and generating pressure, the ventricles propel blood into the pulmonary and systemic circulations. The pulmonary circuit moves deoxygenated blood from the right heart to the lungs for oxygenation and returns it to the left atrium. The systemic circuit distributes oxygenated blood from the left heart to the body, with deoxygenated blood returning to the right atrium through the *venae cavae*. The inferior and superior *venae cavae* deliver oxygen-depleted blood to the right atrium. This blood flows through the tricuspid valve into the right ventricle, which then pumps it through the pulmonary valve and the pulmonary artery to the lungs. Oxygenated blood returns to the left atrium through the pulmonary veins, passes through the mitral valve into the left ventricle, and finally, is ejected through the aortic valve into the aorta. Basically, the blood enters the pulmonary and systemic systems' loops.

The heart's cardiac cycles, a series of pressure shifts enabling blood circulation, ensure oxygen delivery to all body tissues. These cycles are initiated by electrochemical changes in the myocardium, leading to the heart muscle's concentric contraction.

Cardiac cycles consist of two phases: diastole and systole. Diastole, the ventricular filling phase, begins with the closure of the aortic or pulmonary valve and ends with the mitral or tricuspid valve closing, allowing blood to return to the heart.

Systole, the ventricular contraction and ejection phase, starts with the mitral or tricuspid valve closing and ends with the aortic or pulmonary valve closing. During systole, increased ventricular pressure forces blood into adjacent vessels.

3.2.2 Neuro-Autonomic Control of Heartbeat Dynamics

Heartbeat dynamics, among other physiological processes like respiration, digestion, sexual arousal, and blood pressure, is regulated by the complex interactions of the sympathetic and parasympathetic vagal limbs, which are components of the autonomic nervous system (ANS) [160]. The sympathetic nervous system (SNS) is present in any nerve of every living tissue in the body. Its activation may be associated with an elevated activity and attention, the so-called *fight or flight* response, where blood pressure and heart rate may increase. On the other hand, the parasympathetic nervous system (PNS) is smaller than the SNS, and it innervates the head, viscera, and external genitalia. The parasympathetic system, simplistically, boosts the *rest and digest* processes with a lowering of the pressure blood and of the heart rate [88]. Both the sympathetic and parasympathetic nervous systems transmit sensory information to and from the brain and organs. The central nervous system (CNS) functionally controls these peripheral neural systems with input also from the hypothalamus, the limbic system, and the reticular activating system. The nucleus of the solitary tract in the medulla receives sensory signals from the heart and lungs. The connection between the CNS and its target organs involves two types of neurons: preganglionic and postganglionic neurons. Their synapses are located outside the CNS, in autonomic ganglia.

More in detail, the parasympathetic system originates in the brain stem (cranial nerves III, VII, IX, and X) and sacral region of the spinal cord (S_2–S_4). The system governs non-cardiopulmonary processes linked to the rest activity like the production of digestive enzymes for digestion, peristalsis, urination, defecation and salivation, and cardiopulmonary processes including the reduction of the heart and respiratory rates, the blood pressure, and energy preserving through restfulness. The preganglionic and the postganglionic neurons release the acetylcholine, the principal neurotransmitter, which connects to cholinergic receptors. The vagus nerve represents the most important region of the parasympathetic system, because it transmits parasympathetic information to control the physiological functions of the thorax and abdomen structures.

On the other hand, the thoracic and lumbar regions of the spinal cord (T_1–L_2) originate the sympathetic system. It is mainly interested into events requiring alertness, strength, and in which fear, anger, excitement, or embarrassment arises. The sympathetic nervous system increases heart rate, dilates bronchioles for more oxygen intake, and directs blood flow toward the heart and skeletal muscles while reducing blood supply to nonessential organs.

Epinephrine and norepinephrine, released by the adrenal medulla, increase metabolism, stimulate glucose release from the liver, and induce sweating. The sympathetic nervous system also transmits pain signals from organs like the gut, bladder, and uterus.

It is known that the SNS and PNS are complementary, in the sense that a major activity of one normally reduces the role of the other, thus presenting different temporal dynamics, accordingly to their separate activation and physiological functions.

Even though the ANS does not transmit signals to every organ of the body, the nonlinear interactions between the two systems can deliver nonlinear neural information to the targeted organ. The autonomic nervous system's control of the heart exemplifies this, with the sympathetic and parasympathetic branches continuously interacting to modulate cardiac response. This modulation, known as bidirectional augmentation [170], involves a tonic vagal signal increasing the gain of vagal stimulation on heart rate and vice versa. Cytosolic adenosine 3',5'-cyclic monophosphate, crucial to the nonlinear autonomic regulation of heartbeat dynamics, influences this bidirectional enhancement [167].

In other words, the parasympathetic and sympathetic activity effects on heart rate are not purely antagonistic, but they are effect of not simply additive combined sympathetic and vagal stimulation [100], revealing nonlinear and dynamical interactions [167], i.e., simultaneous nervous system changes accordingly to the physiological and pathological stresses in the cardiovascular system.

Heart rate variability (HRV) is influenced by the interplay of sympathetic and parasympathetic activities, along with sinus node function. Physiological factors like blood pressure, temperature, and respiration modulate HRV [186].

3.2.3 Stochasticity in Cardiovascular Dynamics

Since cardiovascular dynamics exhibit both nonlinear and complex behaviors, it is important to discuss the role of noise or stochastic fluctuations and their potential biological origins in cardiac physiology.

While there is no scientific proof of the specific biological dynamics underlying stochasticity in the neurocardiovascular system, related physiological noise likely arises from the nonlinear interaction between the sympathetic and parasympathetic nervous systems. Indeed, autonomic control is the main source of cardiovascular complexity and, consequently, stochasticity [85]. Specifically, a physiological interpretation of cardiovascular dynamical noise could be linked to the production of hormones or neurotransmitters released by components of the ANS, which innervate all body organs and systems. Since the central autonomic network also controls cardiac activity, this stochasticity may influence the dynamics of the brain and body as a whole, including interconnected networks [188, 190].

Norepinephrine, a hormone produced by the sympathetic nervous system, may serve as a physiological source and mediator of neurocardiovascular noise. This hormone may enhance neural control in the prefrontal cortex, particularly during decision-making tasks, by reducing underlying noise [6]. Additionally, norepinephrine system activity appears to correlate with the 1/f noise spectrum of the nervous system [130], suggesting the possibility that neurocardiac noise fluctuations could be modulated by cognitive states.

Furthermore, other chemicals resulting from parasympathetic activity, such as acetylcholine, cytosolic adenosine 3,5-cyclic monophosphate, and other mediators of postjunctional interactions between the sympathetic and vagal systems, influence the combined neural control of heartbeat, potentially contributing to neurocardiovascular noise [167].

3.2.4 Monitoring Cardiovascular Dynamics

Cardiovascular dynamics may be monitored using electrocardiography, which records the heart's electrical activity over multiple cardiac cycles, producing electrocardiogram (ECG or EKG) signals [142]. The ECG is a graph displaying the heart's electrical activity as voltage changes over time, captured by electrodes placed on the skin. These electrodes detect small electrical changes due to the depolarization and repolarization of cardiac muscle during each heartbeat. Abnormal ECG patterns indicate potential cardiac abnormalities.

An ECG is composed of three primary waves: the P wave (atrial depolarization), the QRS complex (ventricular depolarization), and the T wave (ventricular repolarization). During each heartbeat, a healthy heart follows a specific sequence of depolarization, beginning with pacemaker cells in the sinoatrial node, spreading through the atria, passing through the atrioventricular node, and continuing down the bundle of His and Purkinje fibers to the ventricles. This orderly progression creates the characteristic ECG pattern.

Heart Rate Variability (HRV) series, which can be extracted from the ECG, provides valuable insights into the parasympathetic and sympathetic activities of the autonomic nervous system [186]. HRV reflects physiological variations in the time intervals between heartbeats, typically measured as changes in the beat-to-beat interval. It is also referred to as "cycle length variability," "R–R variability" (with R denoting the peak of the QRS complex on the ECG and R–R indicating the interval between successive R peaks), and "heart period variability." The RR interval is used to calculate HRV.

The HRV series is most commonly derived from the ECG (see Fig. 3.1), which is considered the gold standard for HRV measurement because it directly reflects cardiac electrical activity. R-peak identification in the ECG is commonly performed using the Pan–Tompkins algorithm [127]. Exemplary HRV series are depicted in Fig. 3.2.

In certain conditions (e.g., resting state), HRV series can also be derived from photoplethysmogram (PPG) signals, which are optically obtained plethysmograms used to detect blood volume changes in the microvascular bed of tissue. PPG signals are typically acquired using a pulse oximeter, which illuminates the skin and measures changes in light absorption, thereby monitoring blood perfusion in the

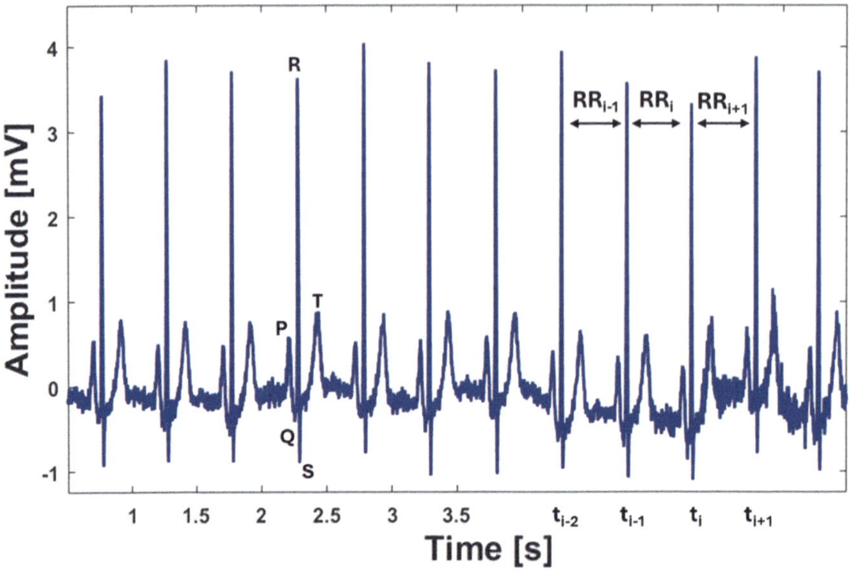

Fig. 3.1 Exemplary ECG signal. The figure represents an exemplary ECG recording, gathered from a subject of the publicly available dataset *Fantasia*, accessible at https://physionet.org/content/fantasia/1.0.0/. On the left part of the figure, the QRS complex is represented, which refers to ventricular contraction, and is preceded by the P wave (atrial contraction) and followed by the T wave (which represents ventricular repolarization). On the right-hand side, it is also showed the formation of the HRV series, namely the time series representing the interbeat time between consecutive R peaks occurring at times t_{i-1} and t_i

dermis and subcutaneous tissues. In addition to monitoring heart rate and cardiac cycles, PPG signals can provide information on respiration, depth of anesthesia, hypo- and hypervolemia, and blood pressure.

3.3 Physiological Noise in the Neural System

3.3.1 Neural System Overview

3.3.1.1 Neurons

Neurons, or nerve cells, are excitable cells and fundamental units of the nervous system that transmit electrical signals called action potentials and chemical signals across a neural network [26]. Neurons are composed of specialized structures [43]:

- *Cell Body (Soma)*: The main body of the neuron, containing the nucleus, mitochondria, and Golgi bodies, responsible for the metabolic functions of the

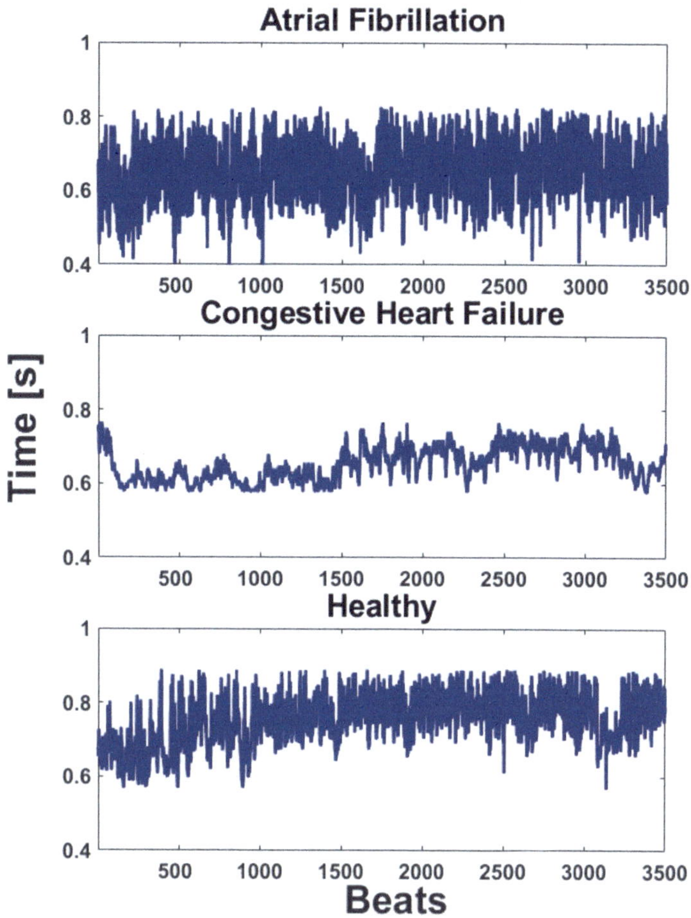

Fig. 3.2 Exemplary HRV series. The figure represents HRV series gathered from a subject with atrial fibrillation condition (upper panel), with congestive heart failure (middle panel), and with no cardiovascular pathology (lower panel). HRV series show the interbeat intervals (y-axis) of the cardiac dynamics for over 3500 beats (x-axis). HRV series are collected at the following links (https://physionet.org/content/nsrdb/1.0.0/, https://physionet.org/content/chf2db/1.0.0/, and https://physionet.org/content/afdb/1.0.0/)

neuron, maintaining the cell, and processing information gathered from other neurons

- *Dendrites*: Tree-like extensions that extend a few hundred micrometers from the soma to collect signals from other neurons. Due to their numerous branches, they can receive input from many other neurons.
- *Axon*: A long, thin projection that extends from the soma at a swelling called the axon hillock. It can stretch over a meter to transmit electrical impulses away from the cell body to other neurons or target cells, such as muscles or glands.

- *Axon Terminals (Synaptic Boutons)*: Located at the farthest tips of the axon's branches, these terminals form junctions with other neurons, muscles, or glands. They release chemical messengers, known as neurotransmitters, which cross the synaptic gap to transmit signals to the next cell.
- *Synapse*: The junction between the axon terminal (presynaptic terminal) of one neuron and the dendrite or cell body of another (postsynaptic terminal), across the synaptic cleft—the gap between neurons
- *Myelin Sheath*: An insulating layer composed of lipids that covers the axon in many neurons. It is formed by glial cells (Schwann cells in the peripheral nervous system and oligodendrocytes in the central nervous system). Its function is to improve transmission along the axon by enabling saltatory conduction, where the action potential jumps between nodes of Ranvier.
- *Nodes of Ranvier*: Small gaps in the myelin sheath along the axon where action potentials are regenerated, enabling rapid signal transmission.

Most neurons receive inputs via the dendrites and soma, while the signal is carried away from the cell body along the axon and passed to other cells through axon terminals and synapses.

3.3.1.2 Glia Cells

Glial cells, also known as neuroglia, are nonneuronal cells in the central and peripheral nervous systems that do not produce electrical impulses [77, 128]. Their main roles are to maintain homeostasis, form myelin in the peripheral nervous system, provide nutrients and oxygen to neurons, ensure insulation between neurons, replace dead neurons, and destroy pathogens. They also play roles in neurotransmission, synaptic connections, and physiological processes such as breathing. In the CNS, glial cells include oligodendrocytes, astrocytes, ependymal cells, and microglia. In the peripheral nervous system, they consist of Schwann cells and satellite cells.

3.3.1.3 Ganglia

Ganglia are clusters of neuron cell bodies, glial cells, and synapses [56]. They are located outside the central nervous system, primarily within the peripheral nervous system. Ganglia function as relay points and processing centers for nerve signal transmission, acting as intermediaries between the central nervous system and peripheral targets, ensuring efficient communication and control over various physiological processes. Ganglia can be categorized into several types based on their location and function, with the two main types being sensory ganglia (also known as dorsal root ganglia or cranial nerve ganglia, which are involved in the transmission of sensory information) and autonomic ganglia (part of the ANS).

3.3.1.4 Central Nervous System

The following description relies mainly on [23, 108, 116, 192]. The CNS consists primarily of the brain, encased in the skull and protected by the cranium, and the spinal cord, which is safeguarded by the vertebrae. The CNS is protected by the meninges, a three-layered barrier made up of the *dura mater*, *arachnoid mater*, and *pia mater*. These layers provide physical protection, guard against neurotoxins, and house cerebrospinal fluid, which cushions the spinal cord. The brain and spinal cord are bathed in cerebrospinal fluid, which replaces the body fluids found outside cells.

The spinal cord is a long, cylindrical structure extending from the base of the brain (medulla oblongata) down through the vertebral column, ending around the first or second lumbar vertebra in adults. It is about 42–45 cm long in adults, with a diameter roughly that of a human finger, though this varies slightly along its length.

The spinal cord is divided into 31 segments, each associated with a pair of spinal nerves, which communicate with the peripheral nervous system (see below).

Located at the center of the spinal cord, gray matter is shaped like a butterfly or the letter "H" in cross section. It contains the cell bodies of neurons, including motor neurons, interneurons, and the initial segments of sensory neurons. The gray matter is divided into dorsal (posterior) horns, which process sensory information, and ventral (anterior) horns, involved in motor control.

Surrounding the gray matter, white matter consists of myelinated nerve fibers (axons) that carry signals up and down the spinal cord. White matter is organized into tracts that transmit sensory information to the brain (ascending tracts) and motor commands from the brain to the body (descending tracts).

The spinal cord performs functions of conduction, reflex actions, and coordination. It serves as the communication conduit between the brain and the peripheral nervous system, relaying motor commands to muscles and glands and transmitting sensory information to the brain. The spinal cord also handles reflexes—automatic, rapid responses to stimuli—without direct brain involvement, allowing for quicker reactions (e.g., the knee-jerk reflex). Additionally, it plays a role in movement and balance coordination.

The brain is the central organ of the nervous system, responsible for receiving sensory information, processing it (including thought, cognition, and intelligence), and coordinating motor control through muscles and the endocrine system. Composed of billions of neurons, the brain's main structural components are the cerebrum, cerebellum, and brainstem.

The cerebrum, which constitutes about 80% of the brain's volume, is the largest part of the brain and is divided into left and right hemispheres, connected by the corpus callosum. Each hemisphere controls the opposite side of the body and is subdivided into four lobes:

- The *Frontal* lobe is involved in executive functions such as reasoning, problem-solving, planning, and voluntary motor activity. The prefrontal cortex within this lobe is crucial for personality expression, decision-making, and moderating

social behavior. The primary motor cortex, also located in the frontal lobe, controls voluntary movements by sending signals to the muscles.
- The *Parietal* lobe is primarily concerned with processing sensory information related to touch, pressure, temperature, and pain. The primary somatosensory cortex, located in the parietal lobe, receives and processes sensory input from various parts of the body, allowing the perception of physical sensations.
- The *Temporal* lobe plays a key role in auditory processing, language comprehension, and memory formation. It contains the hippocampus, essential for the formation of new memories, and the amygdala, which is involved in processing emotions, particularly fear and pleasure.
- The *Occipital* lobe is dedicated to visual processing. The primary visual cortex, located in this lobe, receives and interprets visual information from the eyes, enabling the perception of light, color, and motion.

The cerebellum, located below the cerebrum at the back of the brain, is smaller but essential for coordinating voluntary movements, balance, posture, and motor learning. It integrates sensory input from the body and spinal cord to fine-tune motor actions.

The brainstem connects the cerebrum to the spinal cord and controls basic life functions like heart rate, breathing, and consciousness. It consists of:

- *Midbrain*: It is involved in vision, hearing, motor control, sleep/wake cycles, and alertness.
- *Pons*: It acts as a bridge between various parts of the nervous system, including the cerebrum and cerebellum, and plays a role in breathing and sleep regulation.
- *Medulla Oblongata*: It is the lowest part of the brainstem, controlling vital autonomic functions such as heart rate, blood pressure, and digestion.

Other important structures include the diencephalon, which lies beneath the cerebral hemispheres and contains the thalamus and hypothalamus. The thalamus regulates consciousness, sleep, and alertness, while the hypothalamus controls homeostatic processes. The limbic system, located beneath the cerebrum, is involved in emotions, memory, and behavior. Key structures include the hippocampus (crucial for memory formation and spatial navigation), the amygdala (involved in emotion processing, particularly fear and pleasure), and the cingulate gyrus (regulating emotions and pain and playing a role in predicting and avoiding negative consequences).

3.3.1.5 Cerebral Cortex

The cerebral cortex is the outermost layer of the cerebrum, characterized by its wrinkled appearance due to numerous folds known as gyri (ridges) and sulci (grooves) [117]. This structure plays a crucial role in higher cognitive functions, sensory perception, and motor control. The cerebral cortex is approximately 2–4 mm thick and is composed of gray matter, which primarily consists of neuron cell bodies, dendrites, and unmyelinated axons.

Anatomically, the cerebral cortex is divided into two hemispheres—left and right—connected by the corpus callosum, a thick band of nerve fibers that facilitates communication between the two sides. Each hemisphere is further subdivided into four lobes (frontal, parietal, temporal, and occipital) corresponding to the same cerebrum lobes or areas previously described. Functionally, the cerebral cortex can be categorized into three types of areas: sensory, motor, and association areas [157]. Sensory areas receive and process information from sensory organs, motor areas, such as the primary motor cortex, initiate and control voluntary movements, and association areas integrate information from various parts of the brain, facilitating complex cognitive tasks like learning, memory, and decision-making.

The extensive folding of the cerebral cortex increases its surface area, allowing a greater number of neurons to be packed into the cortical space, thereby enhancing its processing power. This characteristic makes the cerebral cortex the most developed part of the human brain, underlying many of the cognitive abilities that distinguish humans from other species.

3.3.1.6 Peripheral Nervous System (PeNS)

The PeNS [68] is the part of the nervous system composed of nerves and ganglia that lie outside the brain and spinal cord. It is responsible for transmitting information between different areas of the body and the brain. Unlike the CNS, the PeNS is not protected by the skull, vertebral column, or blood–brain barrier.

The PeNS consists of two main divisions: the somatic nervous system, which connects the CNS to muscles, skin, and sensory organs, controlling voluntary movements and transmitting sensory information to the CNS, and the autonomic nervous system (ANS), which regulates involuntary physiological processes such as heart rate, digestion, and respiration.

The somatic nervous system is further divided into sensory and somatosensory systems, which are composed of sensory and somatic nerves.

Somatosensory information is transmitted by 12 cranial nerves in the head and neck. Ten of these nerves originate from the brainstem, primarily controlling anatomical structures in the head. One of them, the vagus nerve, receives sensory signals from the peripheral organs of the thorax and abdomen. The remaining nerve, the accessory nerve, innervates the sternocleidomastoid and trapezius muscles, which are not exclusively located in the head.

Spinal nerves of the PeNS are crucial for carrying somatosensory information throughout the body, except for the head and neck. These nerves originate from the spinal cord and form interconnected networks called plexuses, where nerve roots reorganize into individual nerves that control various body functions. Humans have 31 pairs of spinal nerves: 8 cervical, 12 thoracic, 5 lumbar, 5 sacral, and 1 coccygeal, named based on their location relative to the spinal vertebrae. In the cervical region, nerves exit above their corresponding vertebrae, such as spinal nerve C1, which emerges between the skull and the first cervical vertebra. Below the thoracic level, spinal nerves exit below their corresponding vertebrae.

Cervical spinal nerves C1 through C4 branch out to form nerves in the neck and the posterior part of the head. Spinal nerve C1, also called the suboccipital nerve, controls the muscles at the base of the skull. Sensory and motor functions in the neck are provided by nerves C2 and C3, while sensation at the back of the head is provided by the greater occipital nerve, and the area behind the ears is served by the occipital nerve.

The phrenic nerve, essential for breathing, originates from nerves C3, C4, and C5, and it innervates the thoracic diaphragm. The brachial plexus, formed by nerves C5–C8 and T1, supplies the upper limb and upper back through a complex yet consistently organized network.

The anterior divisions of the lumbar, sacral, and coccygeal nerves combine to form the lumbosacral plexus, with the first lumbar nerve often receiving contributions from the 12th thoracic nerve. This intricate system ensures efficient transmission of sensory and motor information throughout the body via spinal nerves.

3.3.2 *Stochastic Processes in Neural Activity*

According to different neural system scales, there are several physiological sources contributing to the presence of noise in the nervous system.

At the cellular level, neuronal variability can exhibit statistical features such as mean and variance that resemble random processes, though not necessarily being purely stochastic. This variability differs across trials and among individual neurons. While multiple factors influence neuronal trial-to-trial variability, including changes in the internal states of neurons and networks, noise is often considered the primary contributor to this variability [193].

The presence of stochasticity in neurons may account for the variation in the timing of action potentials, even when neurons are driven by identical time-varying stimuli over repeated trials [24, 176].

Possible biochemical and biophysical sources of cellular noise include the functioning of ion pumps, synaptic vesicle fusion, the diffusion and binding of signaling molecules to receptors, and the production and degradation of proteins. Membrane potential fluctuations, even in the absence of synaptic inputs, are influenced by electrical noise, specifically channel noise—random currents caused by the opening and closing of voltage- or ligand-gated ion channels. This channel noise affects action potential timing, propagation along the axon, and trial-to-trial variability [153]. Additionally, noise becomes more significant as neuron size decreases.

Large-scale effects of small biochemical reactions, concentrations, and electrochemical fluctuations in cells may arise from the amplification of nonlinear neuronal activity.

Other sources of electrical noise include shot noise and thermal noise, which are approximately three orders of magnitude smaller than channel noise in central nervous system neurons. Furthermore, *cross-talk* noise occurs due to variations in the activity of nearby neurons, which result from significant changes in extracellular ion concentration following electrical signaling and the spillover of neurotransmitters between unrelated synapses [38].

At the synaptic level, many neocortical cells experience *synaptic background noise*, a multitude of signals from thousands of synapses, which not only contain a meaningful structure on a large scale but also include microscopic sources of true noise from each synapse [42].

Synaptic noise arises spontaneously due to miniature postsynaptic currents, generated by the random release of neurotransmitter vesicles [40]. This randomness is driven by stochastic processes within the synaptic transmission machinery, including the spontaneous release of intracellular Ca^{2+}, noise in synaptic Ca^{2+} channels, random initiation of vesicle-release mechanisms, and spontaneous vesicle fusion with the membrane. Other contributors to synaptic noise include fluctuations in the number of neurotransmitter molecules released per vesicle, random diffusion of these molecules, and noise in synaptic receptor channels, especially when few receptors are present. Receptor protein numbers and densities can also vary randomly over time due to thermodynamic noise affecting protein expression and degradation processes.

In motor and sensory systems, stochasticity is inherently involved through thermodynamic or mechanical noise. In chemical sensing (e.g., smell and taste), noise arises from the random arrival rates of molecules at receptors due to diffusion and the receptors' limited capacity to count signaling molecules accurately. In vision, noise is introduced by the absorption of photons at photoreceptors, governed by a Poisson process, which sets a physical limit on contrast sensitivity, particularly in low-light conditions when fewer photons are available. In movement, noise is produced by interactions with the environment, as well as by human skeletal muscles and muscle fibers [34, 79].

3.3.3 Neural Activity Measurement Techniques

3.3.3.1 Calcium Imaging

Calcium imaging is a microscopy technique used to quantify real-time changes in the concentration of calcium ions ($[Ca^{2+}]$) within living cells [17, 139]. Tracking calcium levels allows researchers to study cellular activity, functioning, and communication among cells.

This technique relies on calcium indicators—fluorescent molecules that react in the presence of Ca^{2+} ions—and on the key role calcium ions play in various cellular processes, such as neurotransmission, muscle contraction, and gene expression. Calcium imaging is primarily used to monitor electrical activity in hundreds of neu-

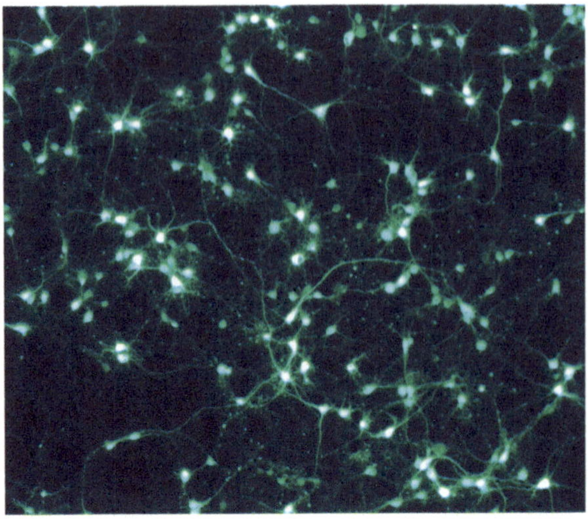

Fig. 3.3 Exemplary calcium imaging. Shown are rat cortical neurons, grown in vitro on a PLL/Lam-coated coverslip, and transfected with a virus to express green fluorescent protein (GFP). This fluorescence picture was taken through a Zeiss Axiovert microscope with the appropriate GFP filter settings (From https://commons.wikimedia.org/wiki/File:GFP_Neurons.png, ManuelSchottdorf, CC BY-SA 4.0 https://creativecommons.org/licenses/by-sa/4.0)

rons in cell cultures or living animals, enabling the observation of neuronal circuit activity during ongoing behavior, since action potential generation is accompanied by a fast internal flow of Ca^{2+} ions. An example of calcium imaging is provided in Fig. 3.3.

The steps in calcium imaging include [57, 139]:

- *Calcium Indicator Choice and Loading*: Calcium-sensitive fluorescent indicators are used, which fluoresce when they bind to calcium ions. These indicators are either chemically or genetically encoded (GECIs) [125]. Chemical indicators are synthetic dyes (e.g., Fura-2, Indo-1, Fluo-3, Fluo-4, and Calcium Green-1) that chelate calcium ions and are typically introduced into cells by permeating the membrane via microelectrodes or micropipettes. GECIs, such as GCaMP, jRGECO1a, and RCaMP1, are genetically engineered proteins that fluoresce when bound to calcium in specific cell types or tissues. These proteins are introduced into cells via genetic engineering techniques like transfection or viral vectors.
- *Cell Stimulation*: Calcium signaling is induced through chemical, electrical, or optogenetic stimuli. Chemical stimuli involve neurotransmitters, hormones, or drugs that affect calcium channels or intracellular calcium stores. Electrical stimuli use electrodes to depolarize the cell membrane, triggering calcium influx. Optogenetic stimuli use light to activate or inhibit calcium channels in cells expressing light-sensitive proteins.
- *Fluorescence Detection*: Upon binding with calcium ions, the indicators emit light at specific wavelengths, and the intensity of fluorescence correlates with calcium concentration. Fluorescence microscopy, using sCMOS or CCD cameras, captures these signals. Confocal, two-photon, and wide-field microscopes can resolve calcium signals in microdomains like dendritic spines or synaptic boutons [121].

- *Imaging and Data Collection*: The dynamic fluorescence changes must be captured with high spatial resolution and for sufficient time duration to observe detailed cellular regions and track the time-varying fluctuations in intracellular calcium levels.
- *Analysis*: Collected data are used to infer quantitative measures of calcium levels, timing, and duration in cells. Fluorescence intensity measurements determine relative calcium concentrations at different time points. Ratiometric imaging calculates the ratio of fluorescence at two different wavelengths to account for variations in dye concentration, providing more accurate measurements of calcium levels. Kinetic and spatial analyses examine calcium signal dynamics, including peak amplitude, rise time, decay time, and signal propagation patterns such as waves or localized bursts.

  From these measures, information can be extracted about neuronal activity, muscle contraction, or signaling pathways.

3.3.3.2 EEG Signals

Electroencephalography (EEG) measures and records the brain's electrical activity in a noninvasive manner, offering real-time reflections of neural dynamics. The following description is based on works by Da Silva [33], Jackson and Bolger [74], Cohen [31].

The genesis of EEG signals begins with the electrical activity of neurons in the brain, particularly the pyramidal neurons located in the cerebral cortex. These neurons, which are aligned perpendicular to the scalp, generate electrical currents through synaptic activity, primarily during excitatory and inhibitory postsynaptic potentials that occur during neuron communication. When groups of neurons fire synchronously, their individual electrical contributions combine to create small electric fields called dipoles. These dipoles produce weak electrical signals that travel through the brain's tissues, cerebrospinal fluid, skull, and scalp, eventually reaching the surface, where they can be detected by electrodes. Figure 3.4 represents a typical device for acquiring EEG signals.

As these electrical signals pass through the brain and skull, they are influenced by volume conduction, a process that not only allows the signals to spread but also attenuates and distorts them. Careful electrode placement on the scalp is crucial to accurately capture brain activity. Standardized systems, such as the 10-20 system, ensure consistent electrode placement across different regions of the scalp to maximize signal detection.

Once detected by the electrodes, these signals are extremely weak, typically in the microvolt range, and require amplification. They are then filtered to remove noise from nonneural sources, such as muscle activity, eye movements, and environmental interferences. This preprocessing step is essential to isolate the neural components of the signals, making them suitable for further analysis.

EEG signals reflect the collective oscillatory activity of the brain, with different frequency bands corresponding to various mental states and brain functions. High

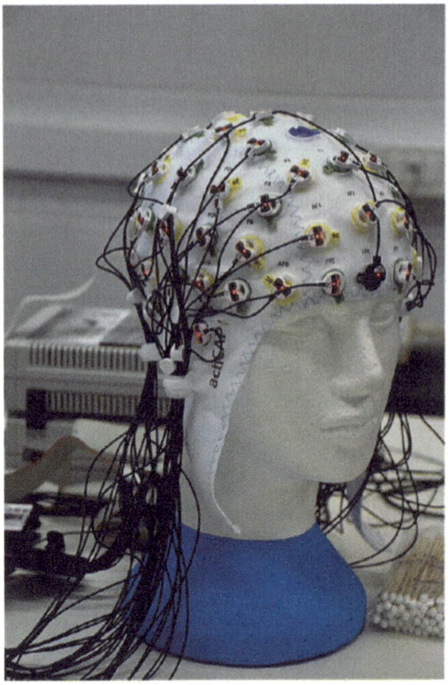

Fig. 3.4 Image of an EEG device placed on the scalp (From https://commons.wikimedia.org/wiki/File:Treinamento_de_EEG-20.jpeg, RIDC NeuroMat, CC BY-SA 4.0, <https://creativecommons.org/licenses/by-sa/4.0>)

sampling rates, typically between 500 Hz and 20 kHz, enable EEG to capture rapid brain activity variations with millisecond-level temporal resolution. For instance, slower waves like delta are prominent during deep sleep, while faster waves such as beta and gamma are associated with active thinking and complex cognitive processing. The real-time capability of EEG to capture these dynamical changes in brain activity makes it a powerful tool for studying brain function in both health and disease, despite its limitations in precisely locating neural events due to the spread of electrical activity through brain tissues. Figure 3.5 shows an exemplary EEG signal.

3.3.3.3 fMRI Signals

Functional Magnetic Resonance Imaging (fMRI) signals arise from the brain's hemodynamic response to neural activity. When neurons in a specific brain region become active, they require more oxygen to support the increased metabolic demands. This increased oxygen demand is met by a corresponding rise in blood flow to that region [161].

The fMRI signal is derived from changes in the magnetic properties of blood, specifically exploiting differences between oxygenated and deoxygenated blood. Oxygenated blood and deoxygenated blood have distinct magnetic properties due to differences in their levels of paramagnetic oxygen. When neurons become active and increase their oxygen consumption, the local increase in oxygenated blood alters the magnetic resonance signal, which can be detected by the MRI scanner [105] (Fig. 3.6).

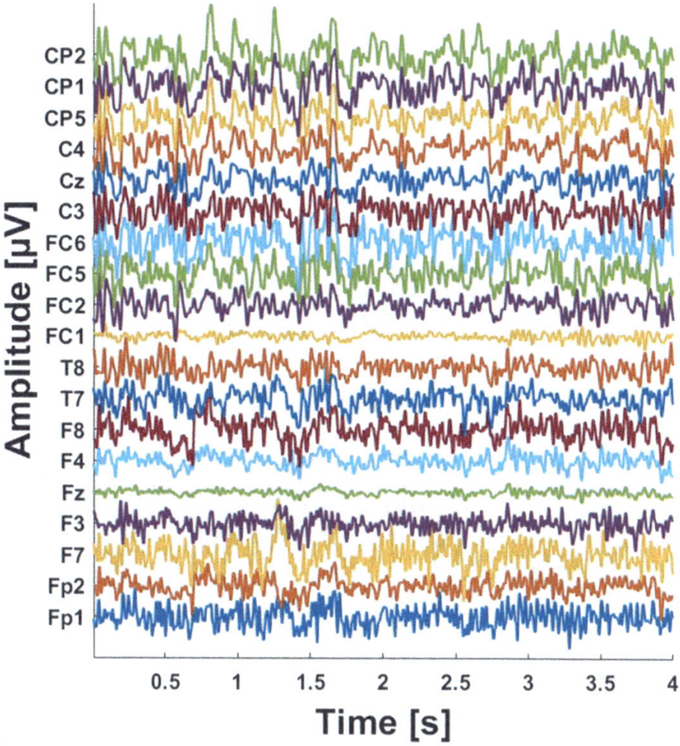

Fig. 3.5 Exemplary EEG signal. The figure shows a collected EEG signal. Each colored line represents the brain activity over 4 seconds for a specified channel on the scalp, reported on the y-axis. The recording has been gathered from a subject of the LEMON dataset described in Chap. 6

Fig. 3.6 Image of an fMRI scanner (From https://commons.wikimedia.org/wiki/File:SNSlab.png EconomicsUZH, CC BY-SA 4.0 https://creativecommons.org/licenses/by-sa/4.0)

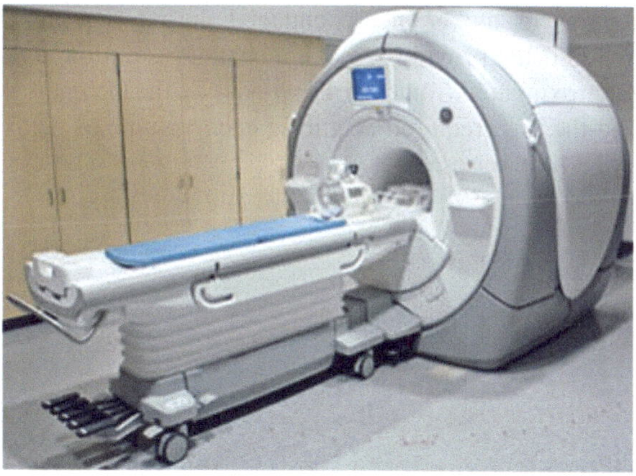

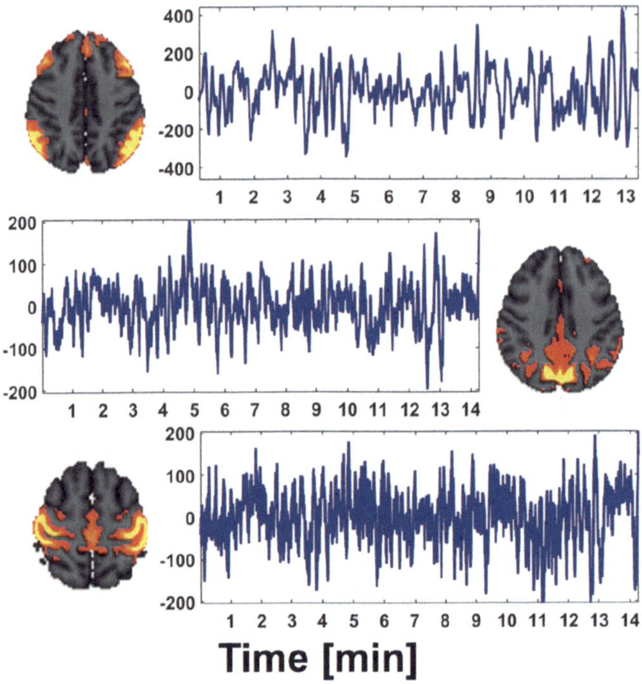

Fig. 3.7 Exemplary fMRI time series. The three panels show exemplary variations of the BOLD activity (y-axis) over 15 minutes (x-axis) gathered from the brain region clusters indicated by the adjacent small heads. The dataset is the HCP1200, described in Chap. 6

In essence, the fMRI signal captures the brain's response to neural activity through the hemodynamic changes that follow, enabling researchers to infer which regions of the brain are engaged during specific tasks or cognitive processes. While fMRI provides excellent spatial resolution, its temporal resolution is limited (approximately from 1 to 4 seconds) by the slowness of the hemodynamic response and the time required acquiring each volume of data. Figure 3.7 shows exemplary fMRI time series.

Chapter 4
Model-Free Approaches for Dynamical Noise Estimation and Reduction

Abstract This chapter describes methodological frameworks for estimating physiological noise in monovariate signals. These approaches are model-free, meaning they do not rely on prior knowledge of the analytical equations governing the underlying physiological processes. The theoretical foundation of the methods is outlined, with a particular focus on the behavior of entropy-based quantifiers under stochastic perturbations. An associated algorithm for the numerical estimation of physiological noise is also provided. The chapter further addresses the application of the methods to signals with time-varying or nonstationary stochastic components, demonstrating its adaptability to realistic physiological scenarios.

The chapter concludes with a brief discussion on the potential for reducing dynamical noise in physiological systems. A simple reduction technique is introduced.

Keywords Physiological noise · Nonstationarity · Denoising

4.1 Monovariate Noise Estimation

In this section a physiological or dynamical noise estimation from a monovariate time series is provided, using the already introduced and discussed *Approximate Entropy* (ApEn) (see Sect. 1.3.4). The latter quantifier is nonlinear, parameter-dependent and developed to detect unpredictability over time series, being considered as an approximation of the Kolmogorov–Sinai entropy for dynamical systems [36], which is actually a theoretical limit value and thus hardly achievable in short-length time series (see Sect. 1.3.1). The proposed noise estimation method particularly focuses on how superimposed noise modifies the ApEn shape, intended as a function of its tolerance or radius parameter.

Electronic Supplementary Material The online version of this chapter (https://doi.org/10.1007/978-3-031-95914-1_4) contains supplementary material, which is available to authorized users.

A. Scarciglia et al., *Physiological Noise and Informative Randomness in Neural and Cardiovascular Systems*, https://doi.org/10.1007/978-3-031-95914-1_4

Indeed, it has been observed that stochasticity or noise presence in a deterministic dynamical system generates an ApEn profile curve with a characteristic behavior, increasing the maximum of the ApEn profile for finite time series, leading to the ApEn divergence for infinite recordings. This concept is resumed in the following result.

The notation $\{a_n\}_n$ is employed to represent an infinite sequence, with a_n being defined as discrete samples arising from a stochastic process or a dynamical system's orbit; conversely, a finite orbit of length $N+1$ is denoted as $\{a_n\}_{n\geq 0}^N$. Moreover, $\{a_n(\upsilon))\}_{n\geq 0}$ is used for denoting the sequence dependence on the generic parameter υ.

Theorem 4.1.1 (ApEn Behavior for Noisy Systems) *Let (X, T, μ) be a discrete dynamical system defined by a C^1 map T, i.e., differentiable, on a compact set $X \subset \mathbb{R}$ and preserving the probability measure μ, whose density h is also C^1. Define the noise as the realization of a stochastic process $\{\varepsilon_n\}_n$ of IID random variables following a distribution q_ξ, described by an unknown parameter ξ. Let $\psi(t)$ denote the probability density function (PDF) of the random variable obtained by taking the difference between any two samples of the noise process. It is assumed that $\psi(t)$ is continuous in a neighborhood of* 0*, and differentiable, at most excluding the value* 0*. For any perturbed orbit of the system $x(\xi) = (x_0, T(x_0) + \varepsilon_1, T(T(x_0) + \varepsilon_1) + \varepsilon_2, \ldots)$, it holds*

$$\mathrm{ApEn}(\{x_n(\xi)\}_{n=1}^{\infty}, m, r) \approx -\log[2\psi(0)r] \tag{4.1}$$

for any embedding dimension $m \in \mathbb{N}$ and for the tolerance $r > 0$ small enough.

It is noticed that the previous result can be weakened to maps T and density functions h that are piece-wise differentiable. This is because the argument used before is not valid for a set of values of z with measure 0.

The result reported in the theorem demonstrates the influence of stochasticity, which causes the quantifier to diverge as the tolerance parameter "sees" the noise effects.

Since the dynamical and proposed physiological noises are thought to be Gaussian to take into account all the possible small independent random contributions present in the system, it is convenient to state the previous theorem for Gaussian distribution of the dynamical noise samples.

Corollary 4.1.2 *If the noise samples are IID Gaussian distributed random variables $\sim \mathcal{N}(0, \sigma^2)$, by setting $\xi = \sigma$, relation 4.1 becomes*

$$\mathrm{ApEn}(\{x_n(\sigma)\}_n, m, r) \approx -\log[r/(\sigma\sqrt{\pi})]$$

and therefore

$$\log(\sigma) \approx \mathrm{ApEn}(\{x_n(\sigma)\}_n, m, r) + \log(r/\sqrt{\pi}) \tag{4.2}$$

when $r < \sigma$ and $r \to 0^+$.

Proof of Theorem 4.1.1 and Corollary 4.1.2 are provided in the Appendix section.

From the corollary, it follows that the noise standard deviation (std) σ can be approximated by the tolerance value r for which the functions $z \to \text{ApEn}(\{x_n(\sigma)\}_n, m, z)$ and $z \to -\log z$ show the most similar differential behavior.

While the proposed noise estimation is theoretically founded for infinite perturbed orbits of dynamical systems, a practical and feasible implementation of the noise estimation in monovariate recording is presented in the next session, while its validity will be tested in the next chapter.

The MATLAB source code for executing the proposed noise power estimation is available at https://github.com/AndScar/noise_estimation.

4.1.1 Noise Estimation Algorithm

Retracing the related theoretical foundation of the framework, it is proposed a noise estimation algorithm relying on the following assumptions:

- The system dynamics is deterministic, taking also into account the possibility of having deterministic chaos, and ruled by any differentiable map T.
- Physiological noise samples are realization of IID Gaussian random variables $\{\varepsilon_n\}_n$ following the distribution $\psi \sim \mathcal{N}(0, \sigma^2)$.
- The system dynamics is perturbed by dynamical noise: For a noise sequence $\{\varepsilon_n\}_n \sim \mathcal{N}(0, \sigma^2)$, any orbit of the dynamical systems is of the form $x_n = T(x_{n-1}, \ldots, x_0) + \varepsilon_n$.

These hypotheses contrast Markovian formalism, in which time series samples are seen uniquely as a sequence of random variables governed by transition probabilities between consecutive states, removing the underlying determinism. Moreover, the properties of the map T appear the most general and suitable to describe and model a wide class of observable phenomena, as long as the systems under study is a dynamical system contaminated by random components.

Now, it is presented the method for estimating the dynamical noise power σ^2 from a noisy series $\{x_n(\sigma)\}_{n=0}^N$ of $N + 1$ samples, using an embedding dimension m and a tolerance step Δr. Its schematic pipeline is presented in Algorithm 1, while its graphical steps have been summarized in Fig. 4.1. The underlying idea is the following: In the first instance, the approximate entropy profile of $\{x_n(\sigma)\}_{n=0}^N$ as a function of the tolerance r is firstly calculated. Then, based on the main result (Theorem 4.1.1), an initial estimate of the noise power $\bar{\sigma}$ it obtained by identifying the value $\bar{r}$ where $\text{ApEn}(\{x_n(\sigma)\}_{n=0}^N, m, r)$ and $-\log r$ have the most similar slope. Finally, the corollary is used to improve $\bar{\sigma}$ by finding the best fit σ for the function $s \to \text{ApEn}(X, m, r) + \log[r/(s\sqrt{\pi})]$ in a neighborhood of $\bar{r}$.

A suitable choice of $I(\bar{r})$ might be $I(\bar{r}) = [r_{\max}, \bar{r}]$, where $r_{\max}$ denotes the tolerance value in which $r \to \text{ApEn}(\{x_n(\sigma)\}_{n=0}^N, m, r)$ reaches its maximum: This interval ensures that the use of corrective terms for the Approximate Entropy (ApEn) [134] is avoided in scenarios where only (or primarily) self-matches occur,

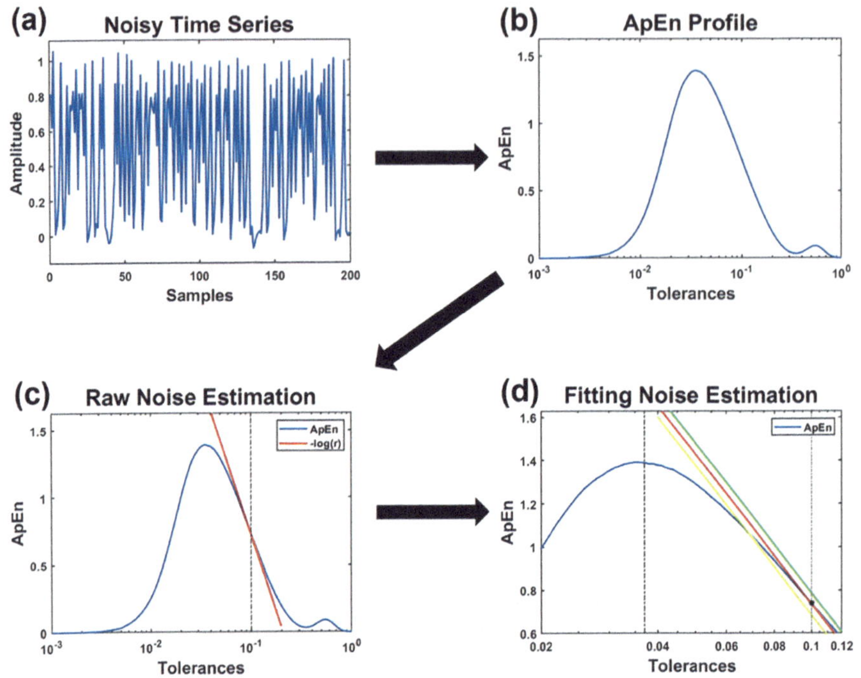

Fig. 4.1 Dynamical noise estimation procedure: take a noisy series as input (panel (**a**)). First, the function $r \to \text{ApEn}(\{x_n(\sigma)\}_{n=0}^{N}, m, r)$ is estimated (panel (**b**))—r ranging between 0 and the series amplitude with a Δr step. A first, raw noise power estimation is performed by searching for the point in which the ApEn profile and the curve $-\log r$ (red line) have the same slope (panel (**c**)). According to the corollary, an improved std estimation is then performed by a curve fitting procedure with respect to the curve $\sigma \to \text{ApEn}(\{x_n(\sigma)\}_{n=0}^{N}, m, r) + \log[r/(\sigma\sqrt{\pi})]$, in the tolerance interval $I(\bar{r}) = [r_{\max}, \bar{r}]$ (delimited by vertical dashed lines), where noise is supposed to be seen (panel (**d**))

Algorithm 1 Noise estimation algorithm

Require $\Delta r > 0$; a noisy series $X := \{x_n(\sigma)\}_{n=0}^{N}$; $m > 0$.
for $r = 0 : \Delta r : (\max(X) - \min(X))$ **do**
  Evaluate $\text{ApEn}(X, m, r)$;
end for
$\text{argmin}_r\ \text{ApEn}(X, m, r) + \log(r) \leftarrow \bar{r}$;
$\bar{r} \leftarrow \sigma$; ▷ Rough Noise Estimation
$\text{argmax}_r\ \text{ApEn}(X, m, r) \leftarrow r_{\max}$;
$I(\bar{r}) := [r_{\max}, \bar{r}]$; ▷ Define a neighborhood of $\bar{r}$
In $I(\bar{r})$, evaluate σ as the best fit of $s \to \text{ApEn}(X, m, r) + \log[r/(s\sqrt{\pi})]$;
return σ

---

particularly when tolerance values range between 0 and the maximum point. In addition, this range ensures that we avoid both the introduction of noise-induced errors and the generation of misleading points with a slope resembling $r \to -\log r$.

For computational efficiency, the ApEn profile across various tolerance values is obtained using the Cumulative Histogram Method (CHM) [179], rather than repeatedly applying the ApEn algorithm [132]. The CHM circumvents the need for multiple executions of the ApEn entropy algorithm, which has a time complexity of $O[n^2]$ for each tolerance r, where n is the time series length. Instead, the CHM calculates the ApEn profile in a single pass, irrespective of the number of tolerance radii, by employing cumulative histograms.

The noise estimation method validation on synthetic data is shown in Chapter 5, using dynamical systems corrupted by dynamical noise. In the following, it is presented a variation of the noise estimation method for stochastic component which is not stationary through the entire time series.

4.2 Nonstationary Physiological Noise Estimation

4.2.1 General Case

Here, a technique for detecting noise variability in a time series is proposed, assuming that the underlying dynamics remains consistent. Following the notation above, the sampled physiological signal or time series $\{x_n(\varepsilon)\}_{n=0}^{N}$ is regarded as an output of a metric dynamical system (X, T, μ), being μ the invariant probability measure defined on the phase space X, on which the differentiable map T defines the deterministic dynamical model. Again, the series is supposed contaminated by dynamical or physiological noise, which can be represented as realizations of a stochastic process $\{\varepsilon_n\}_n$, modeled as independent and identically distributed (IID) Gaussian random variables.

In this context, the nonstationarity hypothesis acts on the dynamical noise power, which is supposed to be time-variant as it follows: By considering s subsets of the sampled series $\{x_n(\varepsilon)\}_{n=0}^{N}$, delimited by subsequent indices $0 = n_1 < n_2 < \cdots < n_S = N$, in which the physiological noise process is stationary, i.e., there exists a finite sequence $\{\varepsilon_{\tilde{N}(t)}\}_{n=n_i}^{n_{i+1}} \sim \mathcal{N}(0, \sigma_i)$, for $i = 1, \ldots, s$.

The method follows this pipeline: For a generic finite noisy time series $\{x_j(\varepsilon)\}_{j=0}^{N}$ with a time-variant noise power, one can select a window W of length $L < N$ and perform monovariate noise estimation (as described above in Sect. 4.1) on the segments of the series from samples x_j to x_{j+L-1}. This yields the quantity σ_j as per [149]

$$\log(\sigma_j|W) \approx \text{ApEn}(\{x\}_j^{j+L-1}, m, r) + \log(r/\sqrt{\pi}) \tag{4.3}$$

or equivalently

$$(\sigma_j|W) \approx \exp(\text{ApEn}(\{x\}_j^{j+L-1}, m, r))r/\sqrt{\pi}, \tag{4.4}$$

for any m, and r sufficiently small, accordingly to Corollary 4.1.2.

In other words, the time-variant noise estimation is achieved by sliding the window W over the entire series, producing a point-wise trend of the noise power, according to the already presented monovariate noise estimation in Sect. 4.1.

4.2.2 Cardiovascular Dynamics

Considering heartbeat dynamics, the previous procedure can be further formalized. Indeed, heartbeat dynamic series are intrinsically discrete and are associated with the R-wave events $\{u_j\}_{j=1}^N$ detected from the ECG, whose samples occur in the continuous-time t. To this extent, $\text{RR}_j = u_j - u_{j-1} > 0$ denotes the jth R–R interval.

Formally, for $t \in (0, T]$, the observation interval, and $0 \leq u_1 < \cdots < u_j < u_{j+1} < \cdots \leq T$ the times of the heartbeat events, let us define

$$N(t) = \max\{j : u_j \leq t\}$$

as the sample path of the associated counting process. Its differential, $dN(t)$, denotes a continuous-time indicator function, where $dN(t) = 1$ when there is an event and $dN(t) = 0$ otherwise. The left continuous sample path is defined as

$$\widetilde{N}(t) = N(t^-) = \lim_{\tau \to t^-} N(\tau) = \max\{j : u_j < t\}$$

such that the cardiovascular system dynamics may be described as

$$RR_{\widetilde{N}(t)} = F(RR_{\widetilde{N}(t)-1}, \ldots, RR_{\widetilde{N}(t=0)}) + \varepsilon_{\widetilde{N}(t)} \tag{4.5}$$

with $\widetilde{N}(t)$ the index of the previous R-wave event before time t and T an unknown, linear or nonlinear function.

The nonstationarity noise component can be represented by considering s subsets of the sampled series $RR_{\widetilde{N}(t)}(\varepsilon(t))$, in which the physiological noise process is stationary, i.e., there exists a finite sequence $\{\varepsilon_{\widetilde{N}(t)}\}_{t=t_j}^{T'} \sim \mathcal{N}(0, \sigma_{\widetilde{N}(t)})$. Therefore, the nonstationary physiological noise standard deviation is evaluated as follows:

$$\psi_{RR}(t', \mathcal{H}_{t'}, \varepsilon(t')) \approx \exp(\text{ApEn}(\{RR_{\widetilde{N}(t)}(\varepsilon(t))\}_{t=0}^{T} W(t-t'), m, r(t')))r(t')/\sqrt{\pi}, \tag{4.6}$$

where $\mathcal{H}_{t'} = (u_{\widetilde{N}(t')}, RR_{\widetilde{N}(t')}, RR_{\widetilde{N}(t'-1)}, \ldots, RR_{\widetilde{N}(t'-T'-1)})$.

Practically, one does not know a priori the temporal intervals where noise achieves stationarity. Thus, it is convenient to select a rectangular time window W comprising w samples and, for each subseries $\{\varepsilon_n\}_{n=i}^{i+w}$, sample-wise estimation of $\psi_{RR}(i, \mathcal{H}_{t'}, \varepsilon(t'))$ is performed.

4.3 Physiological Denoising Method

In this last section, a time-domain method for physiological denoising is illustrated, based on the noise estimation method provided in Sect. 4.1 and on nonlinear forecasting techniques, specifically utilizing Taken's state-space reconstruction [126, 171].

Such a reconstruction has roots in the dynamical system theory and allows for the reconstruction of the multidimensional state space from a univariate time series, creating a *diffeomorphism* that preserves the continuity and differentiability of the starting monodimensional observation and the reconstructed multidimensional dynamics. Importantly, it does not neglect the dynamics of the phenomenon under study. Essentially, when employing state-space reconstruction for forecasting, each corrected point is derived by forecasting using the nearest neighbors in the reconstructed state space from one time step prior.

Again, a finite noisy time series $\{x_n\}_{n=0}^{N}$ is considered with samples $x_n = T(x_{n-1}, x_{n-2}, \ldots, x_0) + \varepsilon_{n-1}$, where T is a generic, unknown linear or nonlinear map defining the deterministic dynamics and $\{\varepsilon_n\}_{n=0}^{N}$ is the physiological noise process, modeled as a realization of a stochastic process of IID random variables with Gaussian distribution $\mathcal{N}(0, \sigma^2)$.

At each time step n, noise irreversibly contaminates the sample x_n, significantly altering the dynamics of the time series or system. To address these perturbations, a time-reversing approach is employed. This method allows for the analysis of how these influences affect the dynamics, as evidenced by the reversed-time order of the original series [172]. Firstly, the time-inverted series is considered $\{z_n\}_n$ such that

$$z_i = x_{N-i+1}$$

for $i = 1, \ldots, N$.

Then, a clean version $\{z_n^{\text{clean}}\}_n$ of the time-reversed series $\{z_n\}_n$ is estimated as follows. A reconstruction of an m-dimensional space is then obtained, generating the embedded series of the form Z_i with $i = 1, \ldots, N - m$, where $Z_i = (z_i, z_{i+1}, \ldots, z_{i+m-1})$. For a fixed index i, the set S_i^{α} of indices is defined such that

$$S_i^{\alpha} = \{j \mid \|Z_j - Z_i\| \leq \alpha\},$$

where the norm $\|\cdot\|$ denotes the infinity or maximum norm in $\mathbb{R}^m$, i.e., $\|x\| = \max_i |x_i|$, for any $x \in \mathbb{R}^m$. The $(i+m)$th time series point is cleaned by substituting it with the mean of all z_{j+m} with $j \in S_i^\alpha$ and such that

$$|z_{i+m} - z_{j+m}| \leq \alpha.$$

It is suggested using $\alpha = 2\bar{\sigma}$, where $\bar{\sigma}$ represents the realization of square root of physiological noise power that, considering the aforementioned assumptions, can be effectively calculated in a closed form as reported in Sect. 4.1:

$$\bar{\sigma} = \exp\{\mathrm{ApEn}(\{x_n\}_{n=1}^N, m, r) + \log(r/\sqrt{\pi})\}$$

when $r < \sigma$ and $r \to 0^+$, for any embedding dimension m. Again ApEn refers to the already introduced nonlinear Approximate Entropy quantifier (Sect. 1.3.4).

The physiologically denoised series $\{x_n^{\mathrm{clean}}\}_n$ is then the reversed-time series of $\{z_n^{\mathrm{clean}}\}_n$.

Heuristically, applying the denoising procedure to the time-reversed series allows for predicting the current state, knowing that its future states are not influenced by noisy effects. In other words, the more reliable value for a time series point is the one that generates orbits close to each other, considering the physiological noise drifting.

All the method validations on synthetic data contaminated by dynamical/physiological noise are reported in the following chapter.

Chapter 5
Synthetic Data Analysis

Abstract This chapter presents numerical simulations that validate the noise estimation methods introduced in the previous chapter. It is organized into sections covering univariate noise estimation, time-varying noise, and physiological noise reduction. The final two sections provide insights into the effects of noise in networks of coupled oscillators and the influence of noise-induced drifts and modulations on complexity quantifiers. For each topic, the relevant datasets, experimental setups, and corresponding results are detailed and discussed.

Keywords Dynamical systems · Izhikevich neuron · Calcium imaging · Synthetic fMRI · Complexity

5.1 Monovariate Noise Estimation

The monovariate noise estimation method is tested on a multitude of maps, gathered from classical analytical benchmarks and from mathematical models generating synthetic physiological signals.

5.1.1 Dynamical Systems Maps

This section presents descriptions of the Logistic and Pomeau–Manneville maps, as well as autoregressive models. The analysis includes consideration of these models' outputs under the influence of dynamical or physiological noise. The experimental setup used for noise estimation is also outlined.

Electronic Supplementary Material The online version of this chapter (https://doi.org/10.1007/978-3-031-95914-1_5) contains supplementary material, which is available to authorized users.

A. Scarciglia et al., *Physiological Noise and Informative Randomness in Neural and Cardiovascular Systems*, https://doi.org/10.1007/978-3-031-95914-1_5

5.1.1.1 Dataset Description

- *Discrete-time logistic maps:* a family of maps described by the equation

$$f_\lambda : [0, 1] \to [0, 1], \quad f_\lambda(x) = \lambda x(1 - x), \tag{5.1}$$

  where λ is a real parameter in the interval $[0, 4]$. The discrete-time logistic map is a simple mathematical model that describes the dynamics of a population over time. It is a nonlinear difference equation that can exhibit a wide range of behaviors, from stable equilibrium points to chaotic oscillations, accordingly to the parameter λ. Simulation is performed with $\lambda = 3.5$, in the periodic regime, and with $\lambda = 3.5699456...$, in the onset of chaos, i.e., in the chaotic regime but with null Kolmogorov–Sinai entropy, and with $\lambda = 4$, i.e., in chaotic regime (Kolmogorov–Sinai entropy > 0).
- *Pomeau–Manneville maps:* a family of chaotic maps described by the function

$$T_\alpha : [0, 1] \to [0, 1], \quad T_\alpha(x) = \{x + x^\alpha\}, \tag{5.2}$$

  where α is a real number greater than 1, and the braces {} denote the reduction modulo 1. For each α, there exists only one T_α-invariant measure μ_α. In addition, the probability measure $\mu_\alpha([0, 1])$ is finite for $1 < \alpha < 2$ and infinite for $\alpha \geq 2$ [111]. Particularly, both Pomeau–Manneville possible scenarios are taken into account, including maps with $\alpha = 1.8$, almost close to the threshold $\alpha = 2$, where the absolutely continuous probability measure, leading is not preserved by the map, and maps with $\alpha = 2.05$, where they do not preserve an absolutely continuous probability measure, leading to a not straightforward application of the Theorem 4.1.1. The latter maps are studied by the results of *infinite ergodic theory*, and it is known that the statistical properties of such maps are peculiar [2, 18].
- *Autoregressive model*: The series of an autoregressive model of order p, denoted as AR(p), can be expressed by the following equation:

$$y_n = \sum_{i=1}^{p} a_i y_{n-i} + \varepsilon_n, \tag{5.3}$$

  where y_n is the n-th observation in the time series, a_i are the p constant coefficients, and ε_n is an IID Gaussian random variable, also known as the innovation process. For the analysis, an AR(7) model has been chosen with the following parameter values: $a_1 = 1.046$, $a_2 = -0.219$, $a_3 = 0.105$, $a_4 = 0.033$, $a_5 = -0.0885$, $a_6 = -0.136$, and $a_7 = 0.105$. These values are obtained by fitting the Yule–Walker equations to an exemplary Heart Rate Variability series of a healthy subject from the NS cohort (the first series, as described in the following chapter). The choice of an autoregressive model of order $p = 7$ is justified by its importance in computational physiology. Autoregressive models enable the characterization of time series regularity in terms of conditional entropy and the variance of the innovation process, particularly when the time series are regarded as Gaussian random variables from an autoregressive model [133].

5.1.1.2 Experimental Setup 1: Noise Estimations on Different Maps

In the following, the method is applied to a plethora of maps. The perturbed series of the maps above are generated as follows:

- Logistic and Pomeau–Manneville: An initial condition is randomly chosen in [0, 1], from which it is produced an orbit by inserting the samples of realization of a white Gaussian process with a given standard deviation at each step of the map equation. Noise standard deviations considered are 0.05, 0.1, 0.15, and 0.2 a.u., which are roughly, respectively, 5%, 10%, 15%, and 20% of the noisy time series range. For logistic maps, it is also included a *bounce* effect: Each time noise moves a series point out of [0, 1], the sample is put back into the interval [0, 1] through reduction modulo 1. This step is necessary for avoiding map divergence.
- AR(7) model: Realizations of the proposed AR(7) model are obtained imposing Gaussian innovation (also known as dynamical noise) with standard deviations of 0.5, 1, 1.5, and 2 a.u.

The length of each series is set to $N = 5000$ samples. For each map, parameter, and noise level, 50 perturbed series with an always different noise realization are obtained; in addition, for any series, noise estimation is performed with a resolution step of $\Delta r = \{\text{series amplitude}\}/1000$, chosen to ensure a reliable estimation of the noise level. Numerical results of the proposed noise estimation method for the above synthetic maps are showed as boxplots in Fig. 5.1. Overall, the method discerns and estimates the different levels of superimposed noise with minimal variance across multiple repetitions. The lower the imposed noise, the lower the dispersion, and hence the better the estimation, even if such a spreading is not large for higher noise values. Even in deterministic chaotic regime (logistic map with $\lambda = 4$ and Pomeau–Manneville maps), the method provides good performances. The same pattern holds true for autoregressive models, where the dynamics are not limited to the interval [0,1]. As the level of noise decreases, the various levels of absolute noise can be identified with greater precision. To provide a sense of the magnitude of the estimated noise levels, it is worth noting that the imposed noise standard deviations are roughly half of the time series' standard deviation for the AR(7) model.

5.1.1.3 Experimental Setup 2: Choice of the Tolerance Step Δr

Here it is demonstrated the importance of selecting an appropriate tolerance step Δr for the noise estimation algorithm used with both synthetic and physiological data. In particular, it is recommended to set $\Delta r = 0.001 \times \{\text{time series range}\}$, where the time series range refers to the difference between the maximum and minimum

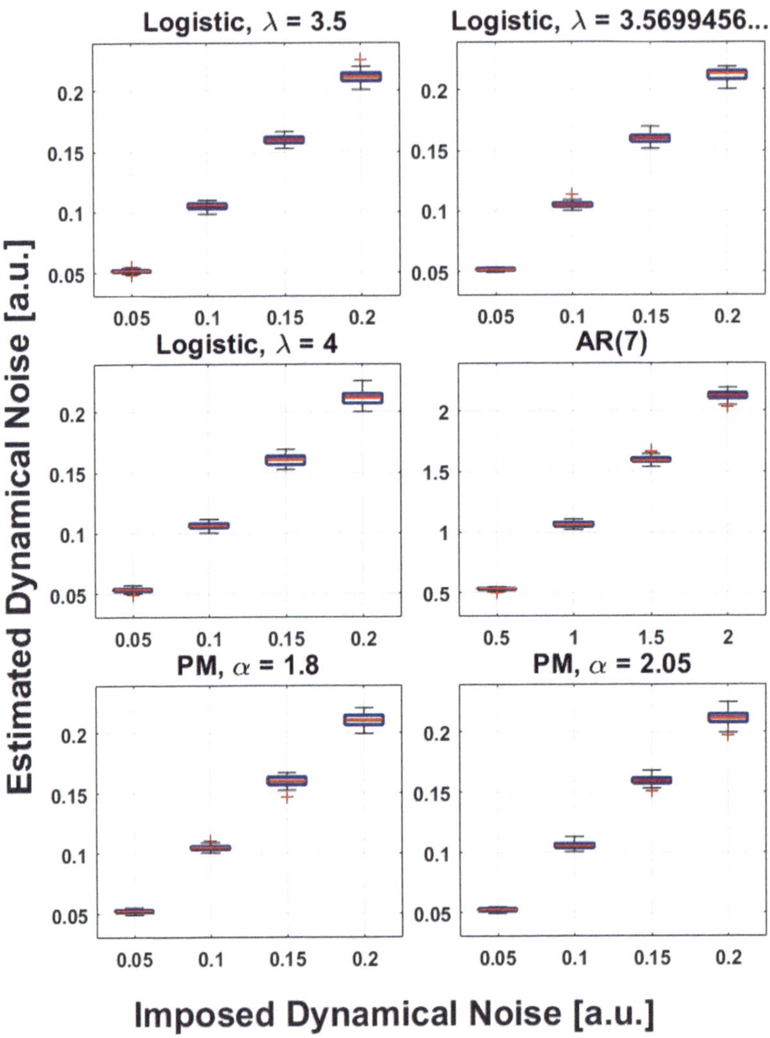

Fig. 5.1 Noise estimates in synthetic data. For each superimposed noise level (x-axis), noise estimations are reported (y-axis), using $m = 2$, $\Delta r = 0.001 \times$ {time series range}, on 50 realizations for each noise level for (top panels) logistic maps in regular regime ($\lambda = 3.5$), in the onset of chaos ($\lambda = 3.5699456...$), (middle panels) chaotic regime ($\lambda = 4$) and for the AR(7) model and for (bottom panels) Pomeau–Manneville maps with $\alpha = 1.8$ and $\alpha = 2.05$. Levels of superimposed and estimated noise are expressed in terms of arbitrary units [a.u.]

values in the series. Choosing the correct value of Δr is crucial as it impacts both the accuracy of the ApEn profile characterization and the computational cost of the algorithm. When a large number of samples are available, a smaller Δr leads to a more precise characterization of the profile. However, physiological recordings

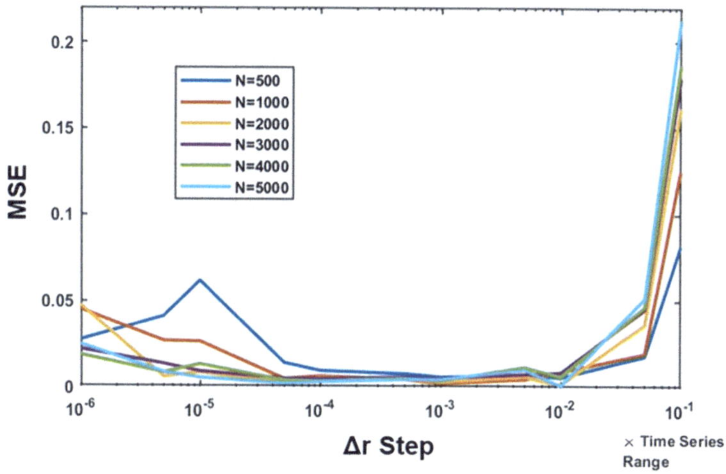

Fig. 5.2 Effect of Δr choice on noise estimation. The performance of the noise estimation algorithm is evaluated with different choices of Δr, ranging from 10^{-6} to 10^{-1} times the range of the time series. Specifically, the mean squared error (MSE) is computed between the estimated noise level (using embedding dimension $m = 2$) and the actual imposed noise for ten different realizations of an AR(7) model with Gaussian innovation standard deviation equal to 1 and series lengths of 500, 1000, 2000, 3000, 4000, and 5000 samples

are often related to short time window and to a limited number of samples. To support the proposal, a series of experiments are performed using ten realizations of the AR(7) model above with Gaussian innovation standard deviation equal to 1. The length N of the time series is varied from 500 to 5000, and the mean squared error (MSE) is computed between the estimated noise and the actual noise imposed with embedding dimension $m = 2$ and different tolerance steps of orders of magnitude $\Delta r = \{10^{-6}, 10^{-5}, 10^{-4}, 10^{-3}, 10^{-2}, 10^{-1}\} \times \{\text{time series range}\}$. Figure 5.2 shows that the MSE is minimized when Δr falls within the range $[0.001, 0.01] \times \{\text{time series range}\}$, independently of the length N of the time series.

Taking into account computational complexity, storage limitations, and machine constraints, it is recommended setting $\Delta r = 0.001 \times \{\text{time series range}\}$.

5.1.1.4 Experimental Setup 3: Effects of the Time Series Length

Since the monovariate noise estimation method of Sect. 4.1 relies on very long observable time series, theoretically infinite, it is worthwhile to ask what happens if one has a limited number of samples. To do so, logistic maps in regular and chaotic regime and AR(7) model above are considered, exemplary representing, maps whose dynamics is bounded and regular, bounded and chaotic, and unbounded and stochastic, respectively. For the logistic maps, noise standard deviations 0.05, 0.1, 0.15, and 0.2 are considered, and noise standard deviations 0.5, 1, 1.5, and 2 are

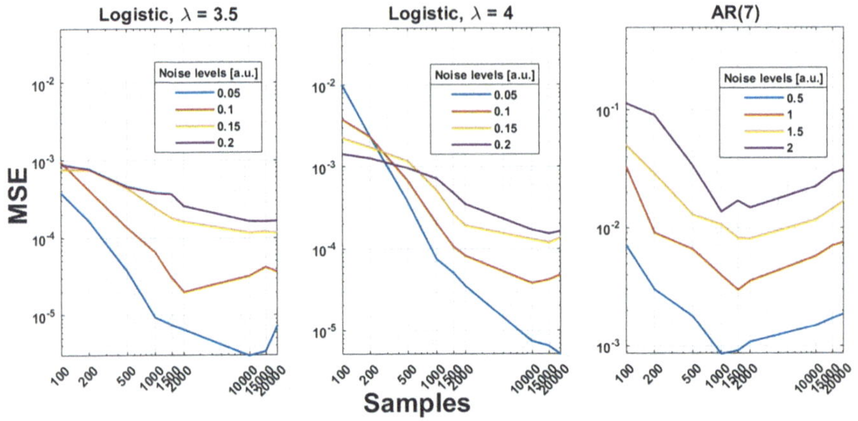

Fig. 5.3 Noise estimates in synthetic data as function of the time series length. For each superimposed noise level (different lines), the MSE of noise estimations and the actual imposed noise are reported (y-axis) as function of the different time series length (x-axis), using $m = 2$, $\Delta r = 0.001 \times$ {time series range}, on 50 realizations for each noise level for logistic maps in regular regime with $\lambda = 3.5$ (left panel), in chaotic regime with $\lambda = 4$ (central panel), and for the AR(7) model (right panel). Levels of superimposed noise are expressed in terms of arbitrary units [a.u.] and reported in the legend

considered for the AR(7) model. For each map and noise level, 50 noisy realizations of 20,000 samples are obtained. Then, noise estimation is applied on the first L samples, with L varying in {100, 200, 500, 1000, 1500, 2000, 10,000, 15,000, 20,000}. The embedding dimension m is set equal to 2, while the Δr step is $0.001 \times$ {time series range}. The behavior of the noise estimation as function of the time series length is reported in Fig. 5.3. In particular, it is depicted the MSE among all the 50 realizations and the actual imposed noise levels, each one represented by a different color. For discrete-time logistic maps, the proposed methodology yields a better noise estimation—the MSE is lower—when a larger number of samples are available, regardless of the noise level considered. For the AR(7) model, the noise estimation improves up to 1000 samples, while it slightly increases for longer lengths, probably due to the nonstationarity of the AR(7) model, for which the theoretical basis of the method does not guarantee the result. In general, lower noise levels are estimated better, as indicated by a lower MSE. In addition, 500–1000 samples seem sufficient for a reliable noise estimation, due to the low MSE reached.

5.1.1.5 Experimental Setup 4: Effects of the Embedding Dimension

Another crucial point in the algorithm is the selection of the appropriate embedding dimension. While, theoretically, the result guarantees a noise estimation independent of the embedding dimension provided infinitely long series are available, it is necessary to observe how this estimation behaves when a finite number of samples

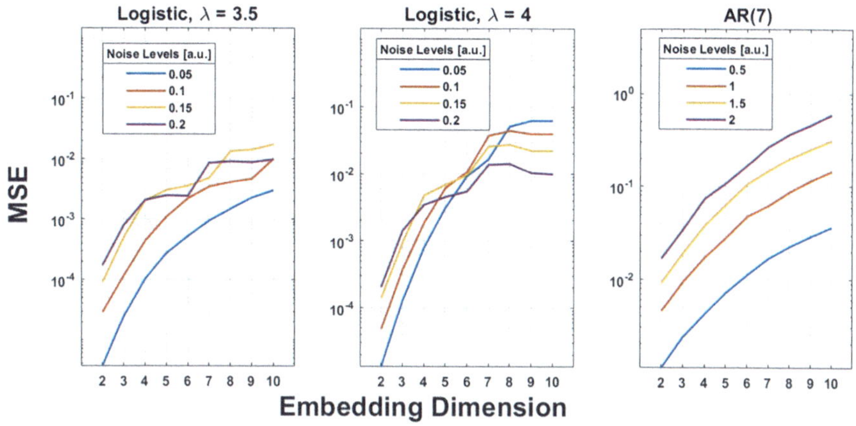

Fig. 5.4 Noise estimates in synthetic data as function of the embedding dimension. For each superimposed noise level (different lines), the MSE of noise estimations and the actual imposed noise are reported (y-axis) as function of the different embedding dimension (x-axis), using $m = 2$, $\Delta r = 0.001 \times$ {time series range}, on 50 realizations for each noise level for: logistic maps in regular regime with $\lambda = 3.5$ (left panel), in chaotic regime with $\lambda = 4$ (central panel), and for the AR(7) model (right panel). Levels of superimposed noise are expressed in terms of arbitrary units [a.u.] and reported in the legend

are available. To do this, we considered, as in the previous case, 50 realizations of 5000 samples of logistic maps and the AR(7) model for the various introduced noise levels. In this case, again for $\Delta r = 0.001 \times$ {time series range}, the noise level was calculated for embedding dimensions m ranging from $2, \ldots, 10$. The results are represented in terms of MSE between the obtained estimates and the actual imposed noise value in Fig. 5.4. It can be seen that, independently of the map and noise level injected, increasing embedding dimension leads to a worse noise estimation, for a fixed number of samples—the MSE increases long with embedding dimension. Anew, lower noises are better found by the method with respect to higher ones.

5.1.1.6 Experimental Setup 5: Sensitivity of the Noise Estimation

Here it is investigated the sensitivity of the proposed method, i.e., its ability to detect linear variations of injected noise. Again, logistic maps in regular and chaotic regimes along with AR(7) model have been considered. In particular, for the former two maps, 30 noisy realizations have been considered for each noise level ranging from 0.01 to 0.2 a.u. with a step of 0.01, and for the latter model, instead, 30 noisy realizations have been taken into account for each noise level ranging from 1.01 to 1.2 a.u. with a step of 0.01. Then noise has been estimated within each series with embedding dimension $m = 2$ and tolerance step Δr as above. Results are depicted in Fig. 5.5, in terms of the median (black line) and of the MAD (shaded area) of the encountered noise levels. Overall, the observed noise levels exhibit a

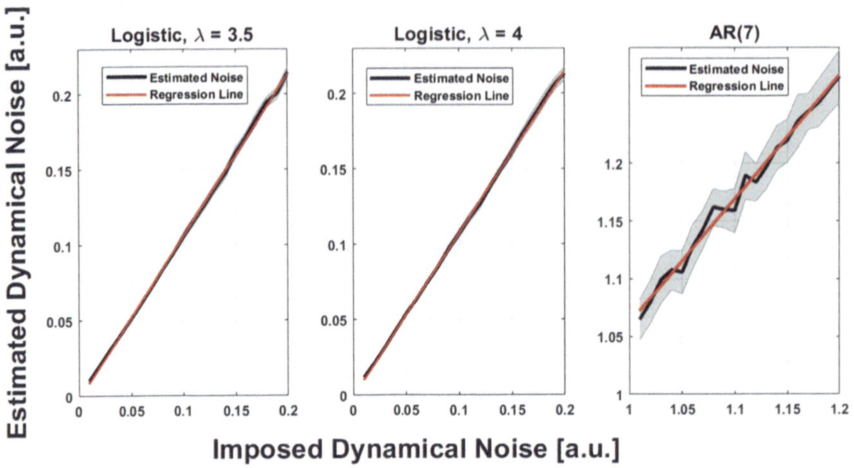

Fig. 5.5 Sensitivity of the noise estimates in synthetic data. For each superimposed noise level (x-axis), the median noise estimations are reported (y-axis) in terms of median (black line) and MAD (shaded area), using $m = 2$, $\Delta r = 0.001 \times$ {time series range}, on 30 realizations for each noise level for: logistic maps in regular regime with $\lambda = 3.5$ (left panel), in chaotic regime with $\lambda = 4$ (central panel), and for the AR(7) model (right panel). The red line denotes the linear regression of the data

linear trend, closely following the imposed one. The variability in noise estimation increases with noise magnitude. Although a slight overestimation is observed, likely due to the finite length of the time series, the linear trend is particularly evident for the logistic maps, as indicated by a significant coefficient of determination ($R^2 \approx 0.99$, $p < 0.05$ according to the F-test), confirming the linear modulation of the injected noise levels. A similar, albeit less pronounced, trend is observed for the noise estimations of the AR(7) model, which align with a regression line yielding $R^2 \approx 0.87$ (F-test $p < 0.05$). This latter behavior may be attributed to nonstationarity conditions, for which the method was not originally designed.

5.1.2 Izhikevich Neuronal Maps

Here it is presented the validation of the monovariate noise estimation on realizations of the Izhikevich neuronal map, a mathematical model used to simulate the electrical behavior of brain neurons, providing a valuable bridge between the complexity of biological neurons and the need for computational efficiency in simulations.

5.1.2.1 Dataset Description

These synthetic data rely on a two-dimensional neuronal model in the form [72]:

$$\begin{cases} v' &= 0.4v \times v + 5v + 140 - u + I + \varepsilon \\ u' &= a(bv - u) \end{cases} \tag{5.4}$$

with the auxiliary after-spike resetting

$$\begin{cases} v &\leftarrow c \\ u &\leftarrow u + d \end{cases} \tag{5.5}$$

if $v \geq 30\,\text{mV}$. Here, the apex denotes the derivative with respect to the time $\partial/\partial t$.

The variables are defined as follows: v represents the membrane potential of the neuron, while u represents the membrane recovery variable—both dimensionless—accounting for the activation of K^+ ionic currents and inactivation of Na^+ ionic currents. The recovery variable provides negative feedback to the membrane potential, v. Following the spike reaching its apex at $+30\,\text{mV}$, both the membrane voltage and the recovery variable are reset. Synaptic or injected direct currents (dc) are delivered via the variable I.

Several dimensionless parameters describe the dynamics of the model. The parameter a determines the time scale of the recovery variable u, with smaller values resulting in slower recovery. A typical value for a is 0.02. The parameter b represents the sensitivity of the recovery variable u to subthreshold fluctuations in the membrane potential v. Higher values of b lead to stronger coupling between v and u, potentially resulting in subthreshold oscillations and low-threshold spiking dynamics. A typical value for b is 0.2. The relationship between b and a can lead to different types of bifurcations. When $b < a$, it corresponds to a saddle-node bifurcation, while $b > a$ corresponds to an Andronov–Hopf bifurcation of the resting state. Finally, the parameters c and d describe the after-spike reset values of the membrane potential (v) and the recovery variable (u), respectively. These resets are caused by fast high-threshold K^+ conductances and slow high-threshold Na^+ and K^+ conductances. Typical values for c and d are $-65\,\text{mV}$ and 2, respectively. Lastly, ε refers to the stochastic noise process that is injected into the dynamics to simulate the dynamical noise effects.

Numerical simulations are conducted to explore different neuronal behaviors using the following parameter settings:

- *Regular tonic spiking*: Parameters $a = 0.02$, $b = 0.2$, $c = -65$, and $d = 6$ are used, with an initial membrane potential $v_0 = -70$ and an initial recovery variable $u_0 = bv_0$. A constant dc of $I = 14$ was injected to induce spiking behavior.
- *Tonic bursting*: Parameters $a = 0.02$, $b = 0.2$, $c = -50$, and $d = 2$ characterized this behavior, with an injected dc of $I = 15$.

- *Mixed mode behavior*: Parameters $a = 0.02$, $b = 0.2$, $c = -55$, and $d = 4$ are used, with an injected dc of $I = 10$.
- *Class 2 excitation*: Parameters $a = 0.2$, $b = 0.26$, $c = -65$, and $d = 0$ defined this behavior. In this case, the injected dc I varied over time.

5.1.2.2 Experimental Setup

Synthetic time series are generated for the aforementioned type of neuronal dynamics comprising $N = 5000$ samples, and then dynamical noise is added. Specifically, such noise is modeled as a Gaussian distribution with zero mean and different standard deviations 2, 5, 10, 15, and 20 a.u. These values roughly correspond to the percentages with respect to the noise-free signal amplitude. For each neuronal behavior and noise level, 100 time series are generated, resulting in a total of 2000 time series. Embedding dimension $m = 2$ and a tolerance step $\Delta_r = 0.001 \times \{\text{series range}\}$ are used to estimate the noise levels within the considered series. This choice of parameters allows to effectively estimate the noise present in the time series data and provides a consistent approach across all models and noise levels.

Exemplary realizations of synthetic neural activity from Izhikevich's model in noise-free and under dynamical perturbation regimes are illustrated in Fig. 5.6.

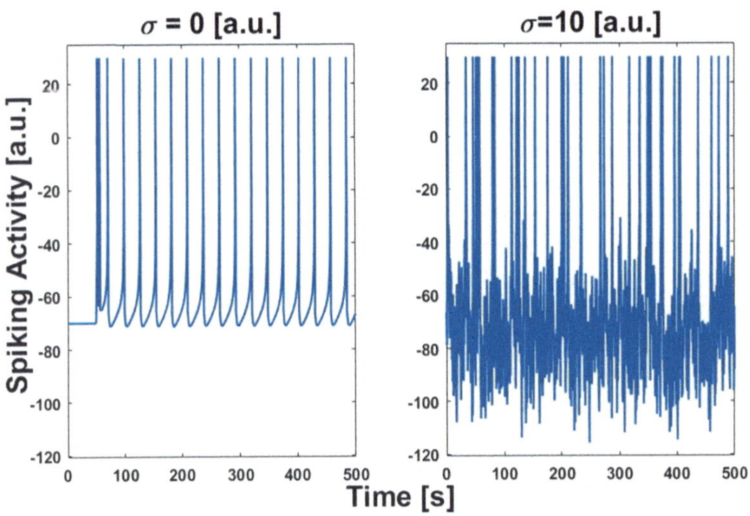

Fig. 5.6 Exemplary realization of regular tonic spiking using the Izhikevich model. The left panel illustrates noise-free spiking activity, while the right panel depicts the same neural dynamics under dynamical perturbation through Gaussian random noise of mean 0 and standard deviation 10

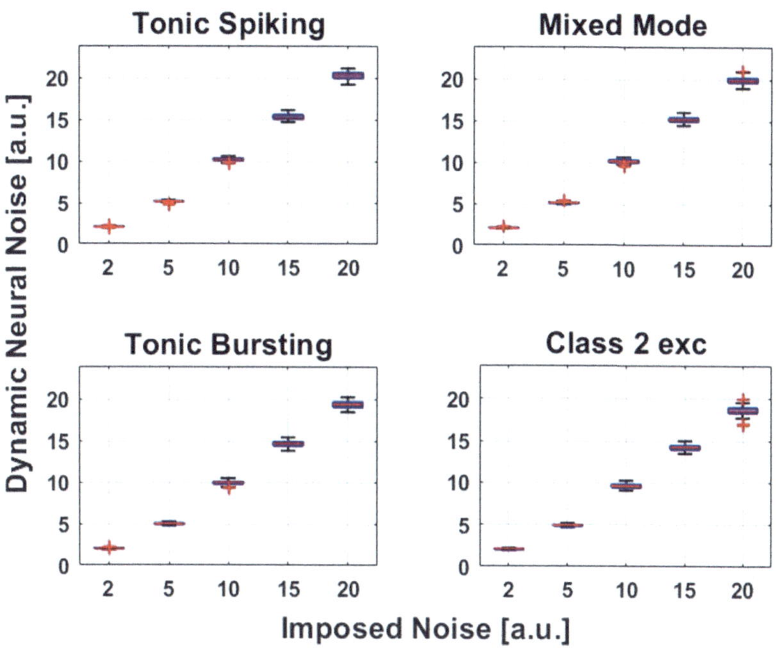

Fig. 5.7 Noise levels estimated for different synthetic neuronal types in the Izhikevich model, including regular spiking, tonic bursting, mixed mode, and class 2 excitation. Each data point on the x-axis represents a specific level of injected noise, and noise estimation was performed on 100 realizations for each noise level. The imposed noise, and hence the dynamic neural noise, refers to the square root of the statistical power of the stochastic realization injected into the series. The noise estimation was conducted using an embedding dimension of $m = 2$, and all time series consisted of 5000 samples

5.1.2.3 Results

Figure 5.7 illustrates the estimated noise obtained for the series generated by the Izhikevich neuronal model using the proposed method. It successfully distinguishes between the different levels, providing an accurate quantification of the underlying dynamical noise.

Overall, the spread around the actual noise value increases as the injected noise level increases. This effect is particularly evident when the imposed noise value corresponds to almost 20% of the amplitude of the noise-free signal.

5.1.3 *Calcium Diffusion Map*

5.1.3.1 Dataset Description

The method in Sect. 4.1 is further validated in estimating dynamic neural noise in a synthetic model for the dynamics of intracellular calcium concentration $[Ca^{2+}]$, described by the following system of equations [102]:

$$\begin{cases} [Ca^{2+}]' = -J_{\text{chan}} - J_{\text{leak}} - J_{\text{pump}} \\ q' = \alpha_q(1 - \text{q}) - \beta_q \text{q}, \end{cases} \tag{5.6}$$

where q represents the fraction of activated IP3 receptor subunits [120], responsible for the generation of the calcium signal. The concentration of calcium ions is controlled by three fluxes: J_{chan}, J_{leak}, and J_{pump}. These fluxes represent the calcium release from the endoplasmic reticulum through IP3 receptors and the calcium influx/efflux from the extracellular space, while α_q and β_q are constant associated with the calcium fluxes. Additional details can be found in [189].

5.1.3.2 Experimental Setup

Time series comprising $N = 50{,}000$ samples with a time step $\Delta t = 0.01$ are generated. To model different calcium dynamics, three IP3 thresholds of saturation are considered: 150, 400, and 1000 (arbitrary units). Such thresholds induced no oscillations, periodic waveforms, and overshoot decaying oscillations approaching a steady value, respectively. On each generated series, the latter $N = 10{,}000$ samples are retained for further analyses. To simulate the existence of dynamical noise in neuronal calcium imaging, Gaussian dynamical noises are incorporated. Specifically, the intrinsic dynamical noise ε is superimposed on the calcium activity, as follows:

$$[Ca^{2+}]' = -J_{\text{chan}} - J_{\text{leak}} - J_{\text{pump}} + \varepsilon. \tag{5.7}$$

The imposed noise levels are 0.5, 1, 1.5, and 2. These values are lower than those used in Izhikevich's maps because the model could not handle higher perturbations without diverging. For each noise level and each IP3 thresholds, 100 time series are generated, for a total of 1500 time series. To estimate the noise levels in these series, embedding dimension of $m = 2$ and the tolerance step as $\Delta r = 0.001$ times the series amplitude are used. This parameter choice allowed us to effectively estimate the noise in the time series data and provided a consistent approach across all models and noise levels.

Examples of synthetic calcium dynamics from the diffusion model in Eq. 5.6 in noise-free and under dynamical perturbation regimes are illustrated in Fig. 5.8.

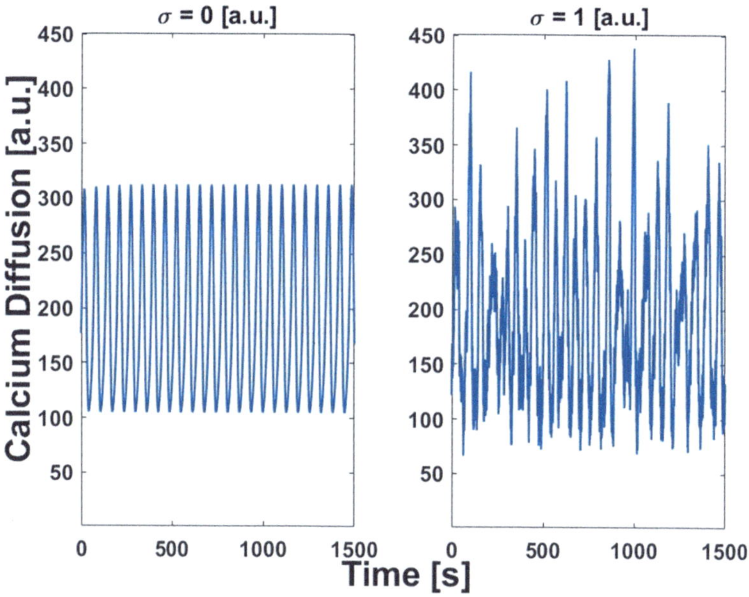

Fig. 5.8 Exemplary realization of calcium dynamics when the IP3 threshold is 400. The left panel illustrates oscillatory calcium dynamics in noise-free condition, while the right panel shows the calcium activity dynamically perturbed with Gaussian random noise of mean 0 and standard deviation 1

5.1.3.3 Results

Regarding the synthetic calcium time series, Fig. 5.9 illustrates a comparative analysis between the estimated dynamical noise and the superimposed ones.

Again, the method successfully distinguishes between the different levels of dynamic neural noise among realizations. Additionally, the method provides an accurate quantification of the various levels of dynamical noise imposed. The noise estimation approach exhibits a narrow dispersion around the median value across the 100 noise realizations for each noise level and spiking behavior. Consistently, the range of variability around the median value expands as the noise introduced becomes more significant.

Thus, results on synthetic data confirm that proposed methodology provides a reliable estimation of dynamic neural noise in nonlinear dynamical systems.

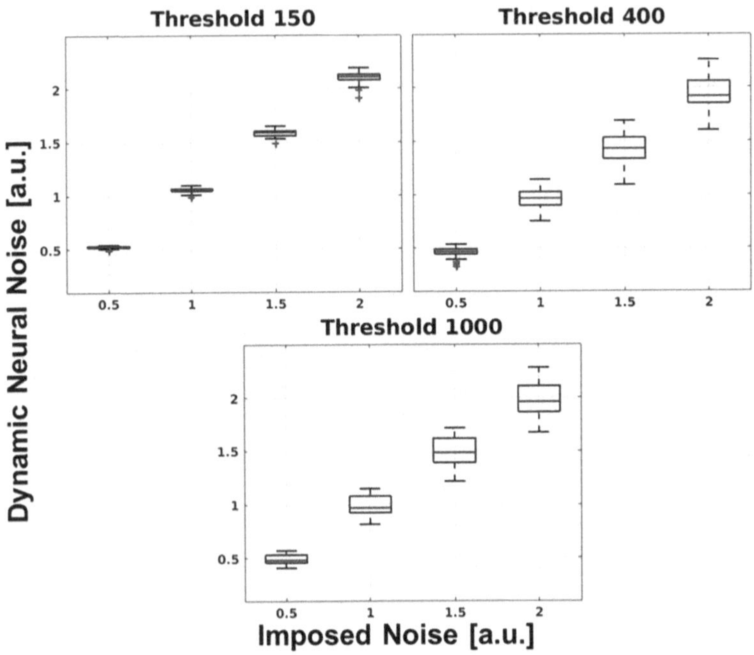

Fig. 5.9 Noise levels estimated for different calcium dynamics for three different IP3 thresholds generating no oscillation, oscillations, and damped calcium oscillations. Each data point on the x-axis represents a specific level of injected noise, and noise estimation is performed on 100 realizations for each noise level. The imposed noise, and hence the dynamic neural noise, refers to the square root of the statistical power of the stochastic realization injected into the series. The noise estimation is conducted using an embedding dimension of $m = 2$, and all time series consist of 10,000 samples

5.1.4 Synthetic fMRI-Like Data

The method for univariate noise estimation has been further validated on noisy synthetic fMRI series, where the fMRI signal is provided by the model described in [48].

5.1.4.1 Dataset Description

According to this model, the fMRI time series is generated by convolving the hemodynamic response function (HRF) with a stimulus, represented as a series of events occurring at random time points. The HRF itself is modeled as the difference of two gamma functions. To introduce the effects of dynamical noise, a sequence of stochastic random variables with a normal distribution $\mathcal{N}(0, \sigma)$ is incorporated into the HRF.

5.1.4.2 Experimental Setup

Two types of simulations on synthetic fMRI data are conducted. In the first simulation (Dataset 1), 100 different noisy fMRI series are generated, each comprising $N = 261$ samples, and time between two consecutive volumes in the fMRI acquisition, i.e., the repetition time (TR), is set at 1970 ms. The series length and TR are selected in accordance with the Cam-CAN dataset (see the next chapter). The dynamical noise is added with a mean of 0 and standard deviation σ, which ranged from $\{0.1, 0.5, 1, 1.5, 2\}$. An exogenous stimulus of 2 s duration is also considered.

In the second simulation set (Dataset 2), the series length and TR are adjusted to 1200 and 0.72 ms, respectively, aiming to replicate the characteristic fMRI series of the HCP-1200 Connectome dataset (see the next chapter). All estimations are performed by setting an embedding dimension of $m = 2$ and a radius resolution of $\Delta_r = 0.001 \times \{\text{time series range}\}$.

5.1.4.3 Results

Results are summarized in the boxplots depicted in the subsequent figures, where the red lines represent the median noise standard deviation observed among realizations, and the lower and upper boundaries of the blue boxes indicate the 25th and 75th percentiles of the distribution.

Indeed, results indicate that the approach effectively discriminates between various levels of superimposed noise; there is a direct correlation between the noise intensity and the dispersion around the median value. Notably, the standard deviation of the HRF superimposed noise aligns closely with the one estimated in the synthetic fMRI signal. Nonetheless, the two datasets exhibit some differences. Specifically, estimated noise values seem to be underestimated in the first dataset (refer to Fig. 5.10c), while in the second dataset, they appear to be more accurate (Fig. 5.10d). It is speculated that these differences are primarily due to the varying series lengths; indeed, the longer the series, the more precise the estimation of the noise levels, as previously discussed in [148]. To address this limitation, separate analyses for each dataset are consistently conducted. It is important to note that the positive trends observed within the two datasets are independent of each other.

5.2 Nonstationary Physiological Noise Estimation

In this section, the proposed sample-wise noise estimation methodology is tested for its ability to detect variations in noise nonstationarity within synthetic time series corrupted by varying levels of dynamical noise. Noise estimation is carried out on a sample-wise basis using overlapping windows of fixed length that slide across the entire time series.

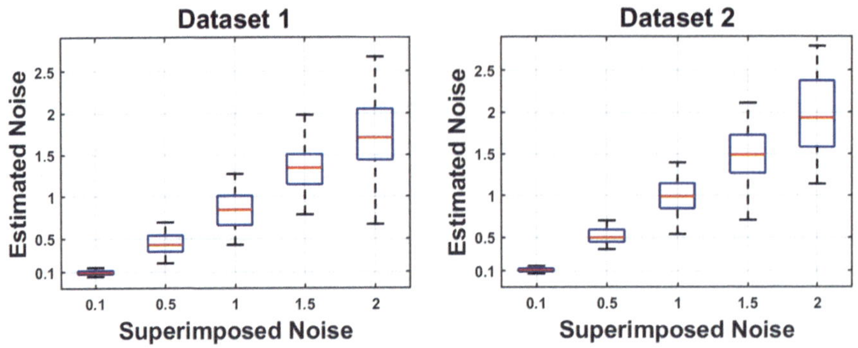

(a) (b)

Fig. 5.10 Noise estimates in synthetic data: For each superimposed noise level (x-axis), noise is estimated using $m = 2$ and $\Delta r = 0.001 \times$ {time series range} on 100 realizations of the noisy synthetic fMRI model. (**a**) Superimposed versus estimated noise in synthetic series from Dataset 1. The series consist of $N = 261$ samples and a TR of 1970 ms, simulating characteristic series of the Cam-CAN dataset. (**b**) Superimposed versus estimated noise in synthetic series from Dataset 2. The series consist of $N = 1200$ samples and a TR of 720 ms, simulating characteristic series of the HCP-1200 dataset

5.2.1 Dataset Description

Synthetic time series are gathered from:

- The already introduced logistic maps (see Eq. 5.1), where the chosen parameters are $\lambda = 3.5$, in periodic regime, and $\lambda = 4$, namely in chaotic regime (Kolmogorov–Sinai entropy > 0). Dynamical noise effects are imposed by adding a noise sample at each step in 5.1.

  For each of the two values of the parameter λ, 50 time series are generated with nonstationary Gaussian dynamical noise.
- Autoregressive models of order p (AR(p)) (see Eq. 5.3 above). The analyses are performed selecting an AR model with order $p = 7$ (AR(7)) with the following parameter values: $a_1 = 1.2710$, $a_2 = -0.4222$, $a_3 = 0.0528$, $a_4 = 0.0544$, $a_5 = -0.0879$, $a_6 = -0.0446$, and $a_7 = 0.1760$. These values are obtained by fitting the Yule–Walker equations to a random exemplary Heart Rate Variability series of a healthy subject from the database https://physionet.org/content/nsrdb/1.0.0/, considering only the first 500 points to avoid nonstationarity. The order $p = 7$ has been selected because in computational physiology cardiovascular dynamics has been modeled through an AR(7) model [133].

  Fifty series are considered for the AR(7) model with nonstationary Gaussian noise.

– Synthetic HRV series: The nonstationary noise detection has been applied also on noisy HRV series generated by the well-known *integral pulse frequency modulation* (IPFM) [21] model. The latter employs a network of interconnected oscillators, with each oscillator representing a distinct aspect of the baroreflex and autonomic nervous system.

  The sympathetic oscillator is characterized by a sinusoidal waveform with a frequency of ω_s and represents the combined low-frequency (LF) power of the HRV spectrum, incorporating vasomotor activity. Conversely, the parasympathetic respiratory oscillator is modeled as a sinusoidal waveform with a frequency of ω_p and aims to capture short-term activity affecting the sinus node through the parasympathetic nervous system.

  The autonomic activity, i.e., the combination of the parasympathetic and sympathetic activities, is represented by the quantity

$$m(t) = C_s s(t) + C_p p(t),$$

  where $s(t)$ and $p(t)$ denote the sinusoidal waveform of the sympathetic and parasympathetic activity, respectively, and where the coupling constants C_s and C_p indicate the degree to which the respective oscillator influences the sinus node oscillator, with s representing sympathetic and p representing parasympathetic modulation.

  The heartbeat series, instead, is generated by integrating an input signal until it reaches a preset threshold of unity. At this point, a pulse is produced, and the integrator is reset to zero. Mathematically, the time t_k of the kth R wave is given by

$$1 = \int_{t_k}^{t_{k+1}} [\mathrm{HR} + m(t)]\mathrm{d}t, \tag{5.8}$$

  where $m(t)$ is the input signal of the autonomic activity and HR represents mean hear rate. Further details on the model can be found in [21].

  Here, $\omega_s = 0.1\,\mathrm{Hz}$, $\omega_p = 0.25\,\mathrm{Hz}$, $\mathrm{C}_s = \mathrm{C}_p = 0.15$, $\mathrm{HR} = 1.1$ with a sampling rate of 250 Hz for the generation of the HRV signals.

  Fifty HRV noisy series of 20,000 samples have been generated. The dynamical noise action is simulated by adding a sample of the noise realization process at the integrand function of each integration step 5.8.

To reproduce the effects of nonstationary noise, for all the previous models, four realizations of the same models consisting of 5000 samples are concatenated, where the ith series is corrupted by dynamical noise as a realization of IID random variables following the distribution $\mathcal{N}(0, \sigma_i(W))$, where $\sigma_1 = 0.05, \sigma_2 = 0.1, \sigma_3 = 0.15, \sigma_4 = 0.2$ for the logistic maps; $\sigma_1 = 0.5, \sigma_2 = 1, \sigma_3 = 1.5, \sigma_4 = 2$ for the AR(7) model; and $\sigma_1 = 0.8, \sigma_2 = 1.2, \sigma_3 = 1.6, \sigma_4 = 2$ for the IPFM HRV model.

To ensure that all samples in the time series remain within the range of [0, 1] without drifting, a modulo 1 reduction is applied at each step of 5.1 for the logistic map.

Eventually, each synthetic series has 20,000 samples formed by four different realizations (5000 points each) where an always different noise realization has been superimposed.

The sample-wise nonstationary noise detection has been applied with different overlapping time window comprising $w = 100$, 150 and 200, with an embedding dimension $m = 2$ and letting the radius r spanning the time series amplitude within the window with a resolution step $\Delta r = 0.001 \times \{\text{time series range}\}$ for all the series.

Moreover, to validate the proposed method's independence from the influence of underlying dynamical systems' behaviors, in contrast to conventional HRV quantifiers, we generated 30 time series by concatenating 3000 points from a logistic map with parameter $\lambda = 3.5$ and 3000 points from a logistic map with parameter $\lambda = 4$, both perturbed with zero-mean Gaussian noise with a standard deviation of 0.05. The two maps exhibit different dynamics: The first is periodic, while the second is chaotic. On these maps we have computed sample-wise STD and noise estimation within a sliding window of 1000 samples.

5.2.2 Results

Figure 5.11 presents the results of sample-wise noise estimation for synthetic data gathered from the logistic maps and the AR(7) model. Each row represents a distinct model, while each column indicates the duration or size of the sliding window used for noise estimation. In each panel, the sample-wise noise estimation (y-axis) is depicted in terms of the median and median absolute deviation (MAD), shown with thick lines and shaded areas, respectively, across 50 realizations within a time window starting from the current point (x-axis) of the time series.

The proposed method effectively distinguishes different levels of nonstationary noise in time series, both qualitatively and quantitatively, even when the sliding window contains a small number of samples. The red horizontal dashed lines correspond to the four variations of injected noise, as discussed in the previous section. Generally, increasing the sliding window size improves noise detection accuracy by reducing dispersion around the median value. The method also identifies transition zones where statistical noise power changes, which are aligned with the steps of the horizontal red dashed lines.

Specifically, higher levels of encountered nonstationary noise exhibit greater variability, indicating the need for a larger number of samples to accurately identify the underlying noisy dynamics. For the logistic map in the periodic regime ($\lambda = 3.5$) and the AR(7) model, the estimated noise values closely match the actual superimposed ones. In contrast, for the logistic map in the chaotic regime ($\lambda = 4$)

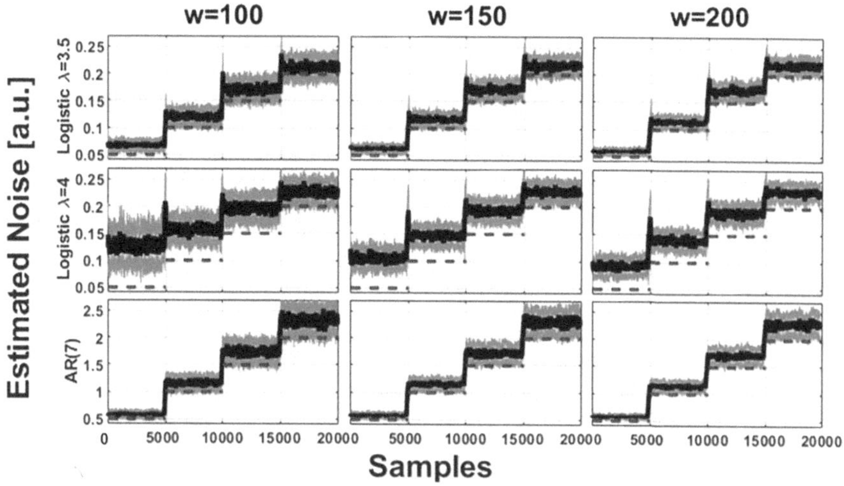

Fig. 5.11 Sample-wise noise estimation for synthetic maps expressed as median (black line) ± median absolute deviation (shaded area) among 50 realizations of the three synthetic data models individuated by the rows. Red lines indicate true reference values. In each panel, the local noise std (y-axis) is illustrated along with the time series length (x-axis). Estimations have been performed with embedding dimension $m = 2$, $\Delta r = 0.001 \times \{\text{time series range}\}$ and within an overlapping time window starting from the sample and comprising w points ahead. Window lengths are indicated by the figure columns

and the AMPA EEG model, there is a slight overestimation that tends to decrease with larger sliding windows.

For the IPFM HRV model, the results in Fig. 5.12 show, across different rows, the mean, standard deviation (STD), root mean square of successive differences (RMSSD), as well as the low-frequency (LF) and high-frequency (HF) powers, and the proposed physiological noise estimation ψ_{RR}. It is evident that while the window length does not significantly affect the estimates, all metrics increase as physiological noise increases, except for the mean values, which remain approximately constant in terms of central tendency.

Importantly, these results demonstrate that changes in time and frequency domain parameters, which are usually linked to sympathovagal and vagal activity changes, may actually be generated by underlying stochastic changes rather than cardiac neural activity. Indeed, such HRV standard estimates do not discern between deterministic and stochastic dynamics.

To further demonstrate this evidence, regarding the possible quantifiers' dependence on the underlying dynamics, results in estimating STD and time-varying noise within overlapping 1000-sample windows of Fig. 5.13 show that STD significantly changes between maps, whereas our proposed method does not exhibit such variability. This aligns with our theoretical claim that our method detects noise as an intrinsic component of the system, independent of the underlying dynamics.

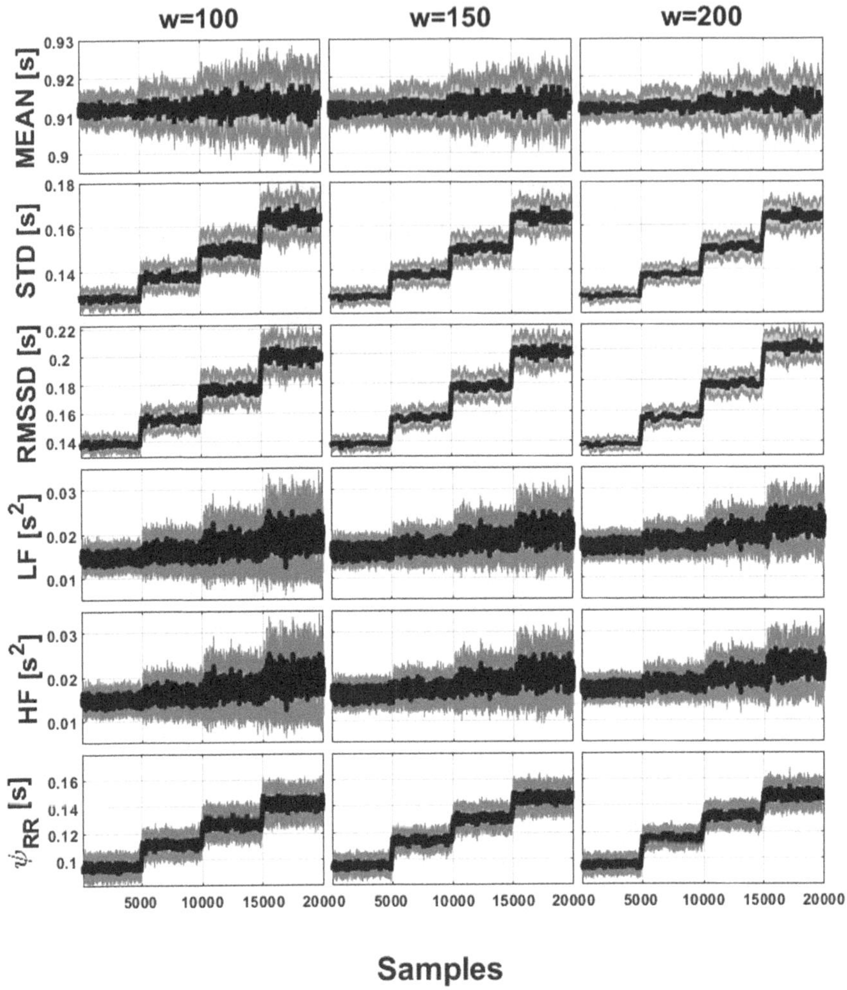

Fig. 5.12 Sample-wise statistics and noise estimation for the IPFM model. Sample-wise estimates of mean, STD, RMSSD, LF power, HF power, and estimated physiological noise are reported on the rows for the IPFM synthetic model, computed across 50 realizations. Each panel illustrates the evolution of these statistics (y-axis) over time (x-axis) within overlapping windows of varying lengths (columns). For LF and HF power estimation, time series were interpolated to 4 Hz. Noise estimation was performed using an embedding dimension (m) of 2 and a radius (r) that spans the time series amplitude within the window, with a resolution step of $\Delta r = 0.001 \times$ {time series range}. The shaded area in each panel represents the median absolute deviation (MAD) around the median value (black line)

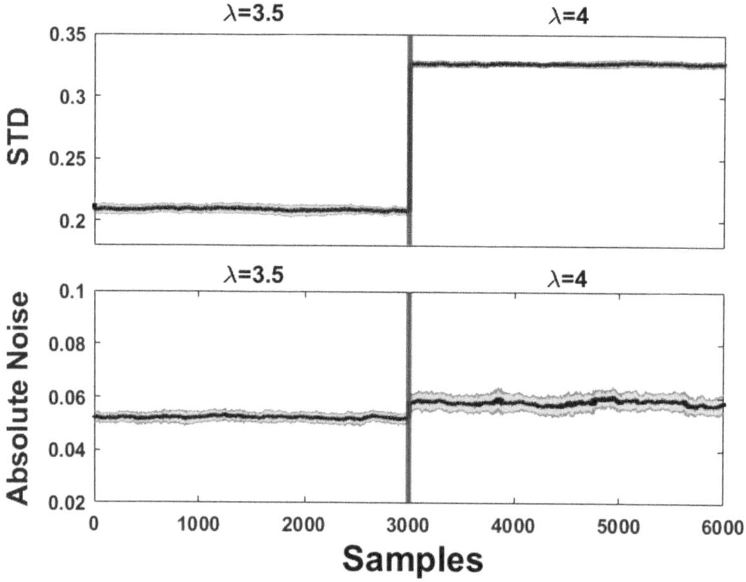

Fig. 5.13 Sample-wise STD and noise estimation for exemplary series with two different dynamics. This figure depicts sample-wise estimates of standard deviation (STD) (top row) and encountered noise (bottom row) for 30 realizations. Each realization consists of 6000 samples, combining 3000 samples from a periodic logistic map ($\lambda = 3.5$) with 3000 samples from a chaotic logistic map ($\lambda = 4$). Both segments are corrupted by the same level of dynamic noise, generated from a normal distribution with mean 0 and standard deviation 0.05. The vertical red line marks the transition between the periodic and chaotic segments. Sample-wise estimates were calculated using overlapping windows of 1000 samples. The black line represents the median value, and the shaded area depicts the median absolute deviation (MAD) across the 30 realizations

5.3 Physiological Denoising Method

The noise reduction method is tested on synthetic time series corrupted by dynamical noise, gathered from the logistic map and the Rössler equations. Unless otherwise specified, the embedding dimension m is set to 3. Moreover, the method's performance has been compared to the closest one in the literature [155]. For this method, parameters $l = 1$ and $k = 1$ are used, resulting in a reconstructed space of dimension 3, according to the notation used in [155]. The performance of the method is evaluated using the dynamical error $E_{dyn}(z_n, T)$, which represents the rooted mean squared error (RMSE) between each sample of the denoised series, z_n, and its deterministic prediction based on past samples, $T(z_{n-1})$, where T is the model generating the map. In other words, the dynamical error quantifies how accurately the corrected series predicts the map iteration from previous values or, equivalently, the extent of error reduction within a single-step iteration.

5.3.1 Dataset Description

The used models are:

- White Gaussian Noise: Realizations of a sequence of IID white Gaussian stochastic random variables are gathered, mimicking purely noisy series. Specifically, considering zero mean, 50 series have been generated for each noise standard deviation amplitude $\sigma = \{0.5, 1\}$ and for the lengths $N = \{2000, 5000\}$, resulting in a total of 200 noisy time series.
- Logistic Map, already introduced above.

  Noisy time series from the model with $\nu = 3.5$ are gathered in a regular regime. Specifically, 50 time series have been collected by taking the first 2000 points from each map realization, starting from random initial conditions in $[0, 1]$, according to (5.1). The standard deviation considered is $\sigma = 0.025$. The same procedure has been repeated for 5000 points and for the other noise standard deviation $\sigma = 0.05$. The introduced noises correspond to signal-to-noise ratio (SNR) levels of approximately 12% and 23%, respectively. Here, the SNR is defined as the ratio of the standard deviation of the noise (σ) to the standard deviation of the signal.
- Rössler Attractor: a set of three nonlinear ordinary differential equations originally proposed by Otto Rössler in 1976. It is one of the simplest systems that exhibit chaotic behavior, making it an important subject of study in chaos theory and nonlinear dynamics. The system is defined by the following differential equations:

  $$\left\{\frac{dx}{dt} = -y - z;\ \frac{dy}{dt} = x + ay;\ \frac{dz}{dt} = b + z(x - c),\right.$$

  where x, y, and z are the state variables representing the system's state, and t is time. Parameters a, b, and c control the behavior of the system. Different parameter values can lead to various behaviors, ranging from periodic orbits to chaotic attractors.

  For numerical simulations, the differential equations are integrated using the one-step forward Euler method with a time step of $h = 0.005$, adding the effects of dynamical noise to the regular Rössler attractor equations:

  $$\begin{cases} x_{n+1} = x_n + h(-y_n - z_n) + \varepsilon_{x,n} \\ y_{n+1} = y_n + h(x_n + ay_n) + \varepsilon_{y,n} \\ z_{n+1} = z_n + h(b + z_n(x_n - c)) + \varepsilon_{z,n}, \end{cases} \tag{5.9}$$

  where $\varepsilon_{x,n}$, $\varepsilon_{y,n}$, and $\varepsilon_{z,n}$ are realizations of Gaussian stochastic processes of mean 0 and standard deviation σ_x, σ_y, and σ_z, respectively. Here n represents the discretized time variable.

  The parameters a, b, and c are equal to 0.1, 0.1, and 6, respectively, exhibiting periodic behavior with orbits of period 2 in the noise-free case. Moreover, σ_y and σ_z are both equal to 0 in order to simulate the effects of the noise perturbation only along the x-axis coordinate.

Fifty synthetic noisy series are generated for each $N = \{2000, 5000\}$ samples and for each $\sigma_x = \{0.05, 0.1\}$. Initial conditions are set randomly and close to the origin. The SNR levels for the two sigma values are approximately 1.5% and 2.5%, respectively. For every noisy attractor, the x-axis for noise reduction is only considered. This process generates a total of 200 noisy series.

The physiological noise reduction method described above is then compared with the most similar denoising technique found in the literature, specifically the method proposed by Schreiber [155], hereinafter referred to as *Schreiber's method*. This approach is model-free, nonlinear, and based on state-space reconstruction. In brief, it embeds the dynamics of a noisy time series into a multidimensional state space and then provides the denoised time series by smoothing the internal points of the embedded vectors that are sufficiently close in the embedded state space. Formally, for two integer parameters l and k, the noisy time series $\{x_n\}_n$ is embedded in $\mathbb{R}^{l+k+1}$, where the embedding vectors are of the following form $X_i = (x_{i-l}, \ldots, x_i, \ldots, x_{i+k-1})$ for $i = l+1, \ldots, N-k+1$. The ith clean sample is then obtained according to

$$x_i^{\text{clean}} = \frac{1}{|\mathcal{U}_i|} \sum_{X_j \in \mathcal{U}_i} x_j,$$

where the values $x_j \in X_j$ are averaged in the neighborhood $\mathcal{U}_i$ which contains all the embedded vectors sufficiently close to the vector X_i. The parameters l, k are arbitrarily chosen, while the size of the neighborhood is determined through coarse-grained partitions. More details can be found in [155].

5.3.2 Results

Figures 5.14, 5.15, and 5.16 illustrate the denoising performance for realizations of white Gaussian noise processes and synthetic Logistic and Rössler maps data presented in the previous section. Panels in different rows correspond to different time series length, and columns denote the different standard deviations of injected dynamical Gaussian noise. In each panel, the root mean square error (RMSE) between the original noise-free series and the denoised ones is represented as a function of the α parameter, namely the amplitude of the neighborhoods used to smooth the series, in terms of median $\pm$ median absolute deviation (MAD) normalized according to the square root of the number of noisy series considered. The lower the RMSE, the better the noise reduction. The red line represents the performance of the method [155] used for comparison, while the blue line denotes the results of the proposed method. The vertical black dotted line refers to twice the median of the encountered noise—computed as in [149]—for the 50 series. In other words, it is a median of the suggested radius to use in the noise estimation algorithm.

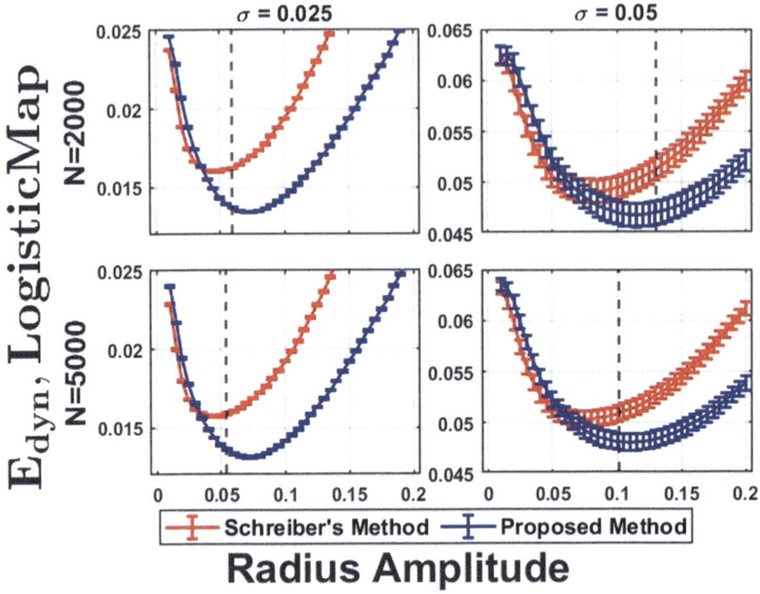

Fig. 5.14 Noise reduction in regular logistic map model. Denoising performance of the proposed method (blue line) and Schreiber's method (red line), in terms of median RMSE (y-axis) ± normalized MAD (with respect to the square root of the number of realizations) between the denoised series and its point prediction among the 50 noisy series realizations, as a function of the neighborhood amplitudes (x-axis). The logistic map is regular with parameter 3.5. The black dotted line represents the median value to use when applying the method in case of unknown dynamics. Each row represents one length of the time series considered, $N = 2000$ and $N = 5000$, while the two columns identify the different levels of superimposed noise

It is evident that the proposed method reduces noise levels better than the existing algorithm for almost any map with deterministic dynamics and that the minimum is usually achieved close to the suggested value $2\bar{\sigma}$. Moreover, it is observed that the lower the imposed noise, the better the noise removal, as less dispersion around the median value is observed. When larger observations are available, i.e., the series is longer, the noise removal is improved for any level of noise, as noticed by the lowering of MAD around the median for the second row of the figure.

While the black dotted line represents a heuristic threshold for optimal noise reduction performance, the actual denoising improvements progressively converge toward this optimal as the number of samples increases. Interestingly, in the case of the logistic map $\sigma = 0.05$, the dynamical error E_{dyn} is close to 0.065 (instead of 0.05), possibly due to the modulo 1 reduction impacting the noisy series generation. For $\sigma = 0.025$, the dynamical error E_{dyn} decreases from 0.025 to 0.0137 and 0.0133 near the suggested threshold, denoted by the black dotted line, for 2000 and 5000 points, respectively. In the other scenario, reductions to 0.048 and 0.047 are observed for 2000 and 5000 samples, respectively.

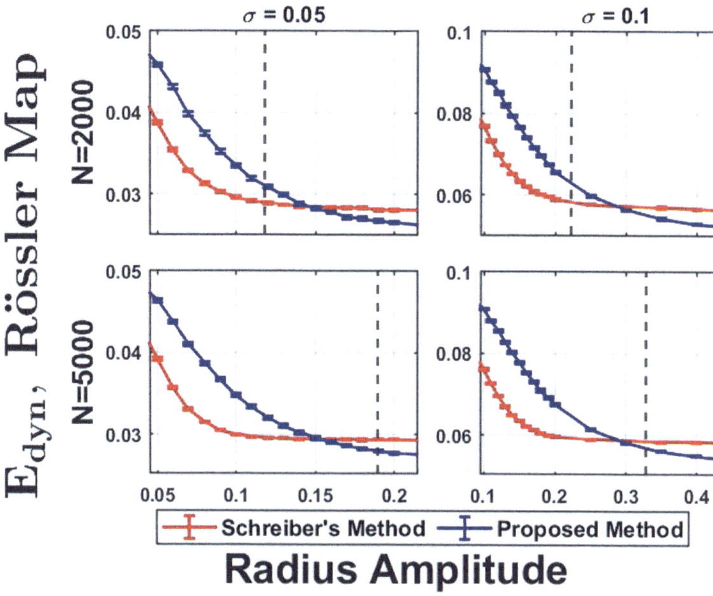

Fig. 5.15 Noise reduction in regular Rössler map model. Denoising performance of the proposed method (blue line) and Schreiber's method (red line), in terms of median RMSE (y-axis) ± normalized MAD (with respect to the square root of the number of realizations) between the denoised series and its point prediction among the 50 noisy series realizations, as a function of the neighborhood amplitudes (x-axis). The Rössler map is regular with parameters $a = 0.1$, $b = 0.1$, and $c = 6$. The black dotted line represents the median value to use when applying the method in case of unknown dynamics. Each row represents one length of the time series considered, $N = 2000$ and $N = 5000$, while the two columns identify the different levels of superimposed noise

For the Rössler map, it has been seen that the suggested median noise level to use in the algorithm increases for longer series, as confirmed by the black dotted line shift toward the right. For $N = 5000$ samples, the dynamical error decreases to 0.028 and 0.057 for the two levels of injected noise, $\sigma = 0.05$ and $\sigma = 0.1$, respectively. Moreover, the suggested threshold appears to be too tight for a more accurate noise removal for the pure Gaussian noise realizations, in which the dynamical error decreases to 0.15 and 0.32 for the scenario involving more samples. Exemplary noise reduction using the proposed method is illustrated in Fig. 5.17 for a realization of White Gaussian Noise (WGN) and a noisy logistic map orbit.

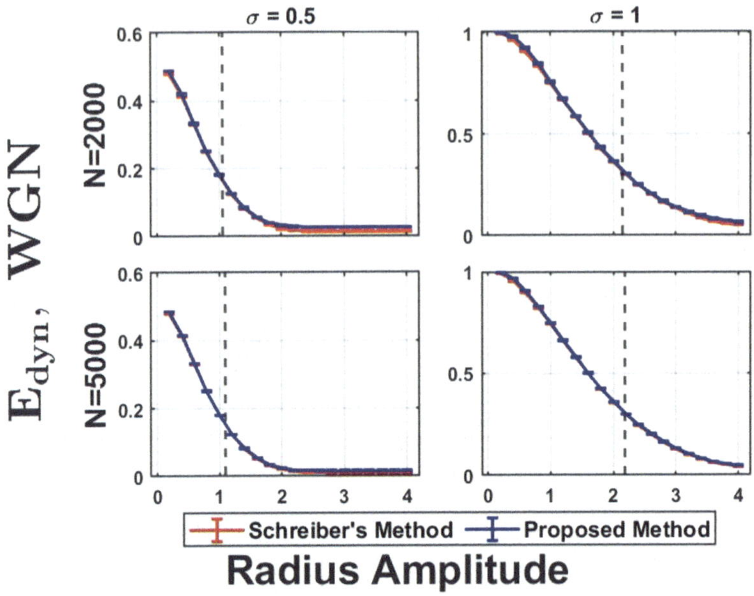

Fig. 5.16 Noise reduction in white Gaussian noise realizations. Denoising performance of the proposed method (blue line) and Screiber's method (red line), in terms of median RMSE (y-axis) $\pm$ normalized MAD (with respect to the square root of the number of realizations) between the denoised series and its point prediction among the 50 noisy series realizations, as a function of the neighborhood amplitudes (x-axis). The black dotted line indicates the median value to use when applying the method in case of unknown dynamics. Each row represents one length of the time series considered, $N = 2000$ and $N = 5000$, while the two columns identify the different levels of superimposed noise

5.4 Dynamical Noise in Coupled Oscillators

Here it is presented a preliminary insight on the effects of dynamical noise in communities of coupled oscillators, considering how the presence of stochastic components can alter the performance of the quantifiers used to describe their qualitative behavior.

The perturbed communities of oscillators are represented by the Kuramoto model, while the complexity quantifiers examined are the integrated information Φ^R and the coalition entropy H_c, which, informally, numerically describe the variability of the system's phases of synchrony.

This analysis is motivated by the use of integrated information Φ^R and coalition entropy H_c in physiological contexts where certain processes are modeled using Kuramoto oscillators. The presence of physiological or dynamical noise could compromise the reliability of the quantifiers, leading to erroneous conclusions in the analysis.

Further details are presented below.

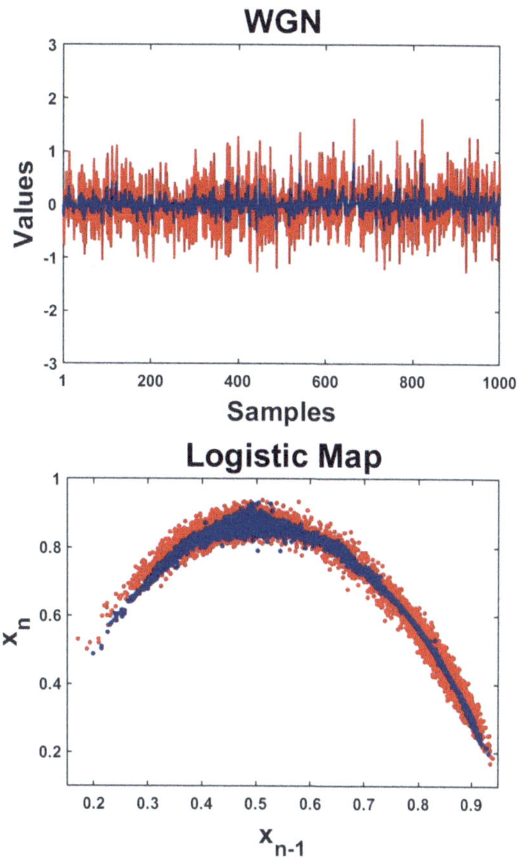

Fig. 5.17 Exemplary noise reduction in a realization of a white Gaussian process and a logistic map perturbed by dynamical noise. The upper panel displays a noise-reduced time series (blue line) performed using the proposed method on a realization of white Gaussian noise (red line) consisting of 1000 samples with a standard deviation of 0.5. The lower panel, on the other hand, depicts the scatter plot of a denoised series (blue circles) obtained from a logistic map with parameter $\mu = 3.5$, which was perturbed by dynamical noise (red circles) with a Gaussian distribution of mean 0 and standard deviation 0.025. For both maps, the embedding dimension is set to $m = 2$. The thresholds used for denoising, respectively, are 1.06 and 0.05, each corresponding to twice the level of noise detected in the series

5.4.1 ΦID

It is supposed that a system X allows for bipartitions, specifically a division into two nonempty subsystems M^1 and M^2, where $X = (M^1, M^2)$. The same principles can be applied to any partition of the space, involving any finite number of nonempty subsets.

According to [10, 114], it is used the integrated Information Decomposition ΦID framework where the fundamental component is a metric known as effective information, denoted as φ^R. This metric assesses the surplus of predictive information, measuring how much more proficient a system X becomes at predicting its own future when regarded as a whole at time τ, as opposed to when viewed as the sum of two subsystems, M^1 and M^2. Essentially, φ^R quantifies the improvement in predictive information generated by the system beyond what the two subsystems can individually produce. For a given bipartition $B = \{M^1, M^2\}$, the effective information of system X beyond B is computed as

$$\varphi^R[X;\tau,B] = I(X_{t-\tau};X_t) - \sum_{k=1}^{2} I(M^k_{t-\tau};M^k_t) + \min_{i,j} I(M^i_{t-\tau};M^j_t), \tag{5.10}$$

where I represents the well-known Shannon mutual information.

The integrated information φ^R linked to the system X across the integration scale τ, Φ^R, satisfies the following:

$$\Phi^R[X;\tau] = \varphi^R[X;\tau,B^{MIB}] \tag{5.11}$$

$$B^{MIB} = \arg_B \min \frac{\varphi^R[X;\tau,B]}{K(B)}, \tag{5.12}$$

$$K(B) = \min\{H(M^1), H(M^2)\}, \tag{5.13}$$

where B^{MIB} is a bipartition of the system X which minimizes φ^R, and K is a normalization constant which reduces the bias in computing Φ^R with unbalanced bipartitions. B^{MIB} is then referred to as the *minimum information bipartition*. Defined in this manner, Φ^R can be interpreted as the minimum information loss experienced when treating the entire system as two distinct subsystems. Practically, the integrated information Φ^R is obtained by computing the information φ^R on all possible bipartitions of the system and by selecting the partition B which minimizes the ratio $B \rightarrow \varphi^R[X;\tau,B]/K(B)$. The quantifier Φ^R is preferred to the classical Φ [10] because the latter can take negative values in case of *redundancy*, and Φ^R controls for redundancy [114, 115].

In essence, interpreting ϕ^R involves various conceptual perspectives that shed light on different aspects of the systems under examination. A particularly relevant interpretation, aligning with complexity science, stems from the theory of information dynamics. This theory breaks down information processing in complex systems into storage, transfer, and modification [104]. Building on earlier findings [114], Φ^R can be conceptualized as capturing a blend of information modification across multiple system parts, information transfer between these parts, and storage in coherent structures spanning more than one system variable.

Alternatively, a more quantitative and mathematically rigorous interpretation of Φ^R revolves around prediction bounds. Conditional entropy, a complement of mutual information, serves as an upper limit on optimal prediction performance. A system with low conditional entropy is predictably accurate. Similarly, higher mutual information between two variables denotes better predictability of a variable from the remaining one. Therefore, Φ^R measures the extent to which the full system state enables superior predictions compared to the states of its individual parts.

5.4.2 Kuramoto Model

Kuramoto oscillators, proposed by Kuramoto [92], serve as a model for the widespread occurrence of coupled oscillators in various natural and engineered environments, attracting considerable scientific attention [131]. While these coupled oscillators are often investigated under conditions that result in the system stabilizing into states of complete synchronization or desynchronization, these extremes fail to fully capture the diverse synchronization phenomena observed in real-world systems. Many systems of interest, such as the human brain, exhibit synchronous rhythmic activity across multiple spatial and temporal scales without settling into stable states.

To address this challenge, an extension of the Kuramoto model introduced by Shanahan is employed [156], which incorporates the occurrence of synchrony transitions.

This network consists of N communities, each containing m oscillators. Every oscillator is connected to all others within its community and has a probability q of being coupled to each oscillator in the rest of the network. The state of the ith oscillator is determined by its phase θ_i, governed by the evolution equation:

$$\frac{\mathrm{d}\theta_i}{\mathrm{d}t} = \omega + \frac{1}{k}\sum_j K_{ij}\sin(\theta_j(t) - \theta_i(t) - \beta - \pi/2) + \varepsilon_i(t), \tag{5.14}$$

where ω is the natural frequency, β is a constant tuning the phase lag, K_{ij} are the connectivity coefficients, k is a normalization constant, and ε_i are noisy drifts. The coupling between oscillators i and j is defined based on the community structure:

$$K_{ij} = \begin{cases} a & \text{if } i \text{ and } j \text{ are in the same community,} \\ b & \text{otherwise,} \end{cases} \tag{5.15}$$

where $a > b$. Essentially, K is an $N \times m$ block matrix where the square blocks of size $m \times m$ along the diagonal consist entirely of entries with a value of a. Each row of K outside these diagonal blocks has all elements set to 0, except for $q \times m \times (N-1)$ randomly positioned values, which are assigned a value of b.

The dynamical properties of the oscillators are assessed by the instantaneous synchronization R. The instantaneous synchronization at time t of community c in the set $\{1, \dots, N\}$, comprising oscillators I_c, quantifies their concentration and aggregation in the θ-space,

$$R_c(t) = |\langle \mathrm{e}^{i\theta_j(t)} \rangle_{j \in I_c}|. \tag{5.16}$$

For simplicity, Φ^R is computed with respect to the coalition configuration of the system, which is defined for each community c and time t as

$$X_c(t) = \begin{cases} 1 & \text{if } R_c(t) > \gamma, \\ 0 & \text{otherwise,} \end{cases} \tag{5.17}$$

where γ represents a binary threshold on the continuous level of synchrony for considering a community to be synchronized or not. This representation offers N interdependent binary time series $X_t := (X_t^1, \ldots, X_t^N)$, indicating the set of communities that are more internally synchronized. Considered as an N-dimensional time series evolving over time, the coalition configuration $\{X_t\}_t$ depicts the progression of synchrony configurations. The diversity in synchronization patterns across communities can be measured using the Shannon entropy over time, referred to as coalition entropy H_c. It essentially represents the uncertainty in N-dimensional time series vectors that describe synchrony configurations. This metric is associated with the amplitude coalition entropy measure in [150].

It has been shown in [156] that the communities generated by the Kuramoto equations are completely desynchronized or hypersynchronized, for tuning parameter β close to 0 and above 0.3, respectively. Metastable behavior—i.e., communities undergoing fluctuations between states of high and low synchrony—is obtained when β lies between these two values.

5.4.3 Experimental Setup

Simulations are conducted on a network consisting of $N = 8$ communities, each comprising $m = 32$ oscillators. A connection strength of $a = 0.6$ is imposed inside each community, while it is $b = 0.4$ across different communities. An oscillator can be connected to any other oscillator not belonging to the same community with a probability of $1/8$. The natural frequency is set to $\omega = 1$.

It is assumed that the realizations of the noise processes, denoted by $\{\varepsilon_i\}$, are collected from independently and identically distributed (IID) normal variables with a mean of 0 and a standard deviation of σ. For each level of $\beta \in \{0, 0.1, 0.3\}$ and for the noise standard deviation $\sigma \in \{0, 0.05, 0.1, 0.15, 0.2\}$, 50 simulations of perturbed orbits are for the Kuramoto model, using a single-step forward, explicit Euler algorithm with a step size of 0.05 for numerical integration—the outcomes in synchrony, under noiseless conditions, exhibit no variance compared to [115], which employed a different integration method.

The three values of β denote complete asynchrony among the communities, metastable chimera states, and complete synchronization. At each trial, the initial conditions in Eq. 5.14 are chosen arbitrarily in $(0, 2\pi)$, while the realization of the stochastic sequence $\{\varepsilon_n\}_n$ is supposed to be IID Gaussian $\mathcal{N}(0, \sigma)$ and always different for any fixed noise level σ. For each simulation, the first 10^4 samples are discarded to avoid transient effects on the dynamics, obtaining series of 10^5 samples, simulating a total observational time of 5000 s $\approx$ 1.4 h. According to [115], the threshold γ for defining the coalition is set equal to 0.8. All information-theoretic measures are reported in bits.

Once the noisy trajectories have been generated, the coalition configuration is first obtained according to 5.16 and 5.17, and then both the integrated information Φ^R, following formulas 5.11–5.13, and the coalition entropy H_c are computed. In order to assess the relationship between ΦID measures and the different series lengths, the measures have been computed for truncated series containing the following numbers of samples: 100, 250, 500, 1000, 2500, 5000, 8000, 10,000, 15,000, 20,000, 25,000, 50,000, 75,000, and 100,000. Approximately, the previous values correspond to dynamics describing from one up to 1000 natural oscillations. Φ^R has also been computed for the time lags $\tau = 1, 10$, and 100, measured in integration time steps.

5.4.4 Results and Remarks

Figure 5.18 illustrates how the integrated information measure Φ^R changes with the length of the time series for different levels of injected noise (rows) and various phase lags (columns). It is stressed that the amount of dynamical noise introduced into the system is precisely determined for each oscillator series, wrapped on $[0, 2\pi]$ by using the monovariate noise estimation method described in Sect. 4.1.

Upon initial observation, significant trends emerge with large time delays ($\tau = 100$). It is apparent that the effects of noise do not alter the qualitative behavior of the networks when they are fully desynchronized ($\beta = 0$). On the other hand, under conditions of high synchrony ($\beta = 0.3$), noise significantly impacts the overall coupling, causing disruptions in synchrony through stochastic changes in dynamics. Interestingly, the metastable chimera state ($\beta = 0.1$) appears highly susceptible to noise injection, resulting in a consistent reduction of Φ^R with increasing noise levels. However, regarding the series length, except for complete asynchrony, the measure profile exhibits a peak for large time delays and loses variability around the median value for longer observational times.

Additionally, in Fig. 5.19, the coalition entropy is seen to mirror the behavior of the Φ^R measure: It remains unaffected by noise when the communities are completely desynchronized; it increases when noise is injected in highly synchronized coupled communities. The median entropy and its variability are reduced only in the metastable case. Furthermore, the coalition entropy consistently increases with the series length, while the variability around the median value typically diminishes, though this is not clearly evident for the metastable state with small amounts of noise.

The measure Φ^R is tested on Kuramoto's equations, which analytically represent systems of coupled, interacting oscillators. It has been established how the measure

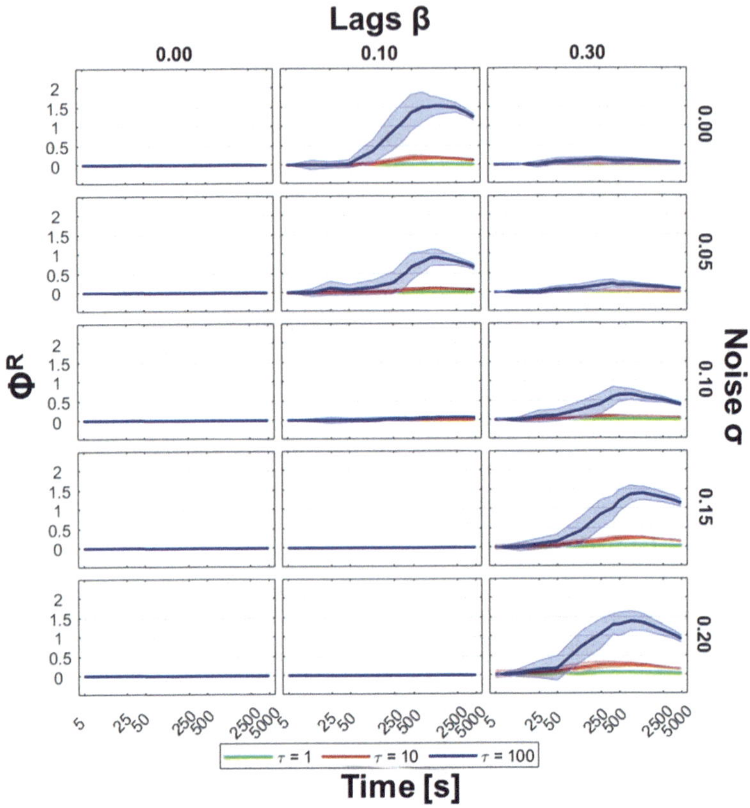

Fig. 5.18 Φ^R measure for noisy networks of Kuramoto oscillators. The panels depict the behavior of the integrated information measure Φ^R across different phase lags β (columns) and increasing noise levels (rows). Each panel shows Φ^R in terms of the median (thick lines) and the mean absolute deviation (shaded area) among 50 different noisy networks. The x-axis represents the observation time, while the y-axis displays the Φ^R values. The colors green, red, and blue represent the time delays $\tau = 1, 10, 100$, respectively

Φ^R behaves with observation time and in dynamical noise regimes for three types of global state: complete asynchrony, metastable chimera, and hypersynchronization.

The decreased value of Φ^R at short time lags could stem from redundancy domination, where the limited time span impedes extensive system-wide interactions. Conversely, increased Φ^R values for longer time lags are associated with overall system synergy, where the system's predictive ability is facilitated by the interaction between its individual components [115].

Noise plays a crucial role in information characterization, especially for the peculiar metastable chimera and hypersynchronization states. Indeed, the complete asynchrony ($\beta = 0$) is accentuated by the drift of the added noise and leads to $\Phi^R = 0$. In contrast, the behavior of closely synchronized coupled networks

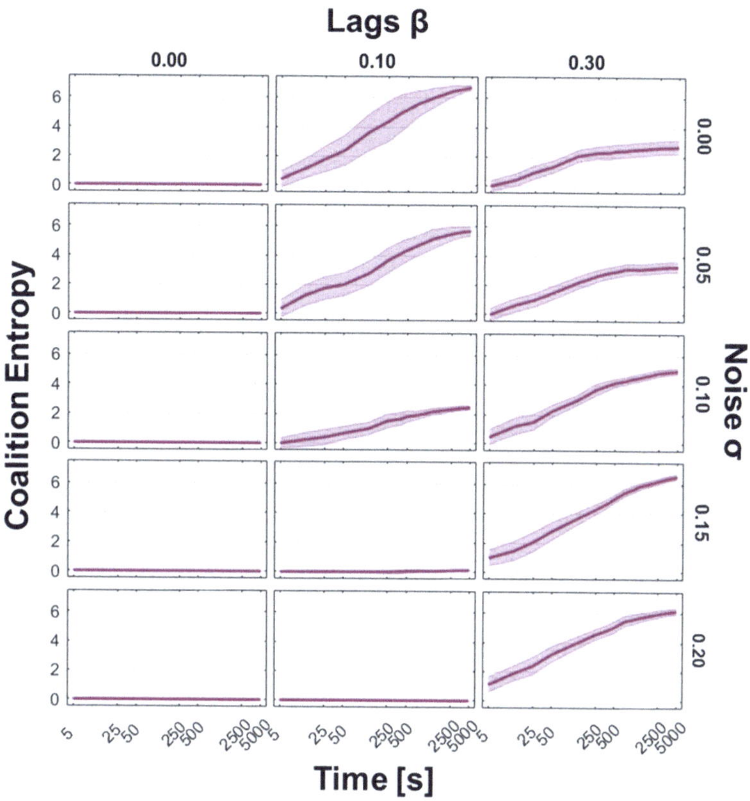

Fig. 5.19 Coalition entropy for noisy networks of Kuramoto oscillators. The panels represent the coalition entropy for different phase lags β (columns) and for increasing injected noise levels (rows). In each panel, the coalition entropy is represented in terms of median (thick lines) and the mean absolute deviation (shaded area) among 50 different noisy networks. The x-axis represents the observation time, while the y-axis displays the coalition entropy values

($\beta = 0.3$) experiences substantial impact from stochastic perturbations and the accompanying noise powers. These perturbations induce the system to depart from complete synchrony, leading it into states of apparent synchronized intermittency. For what concerns the metastable chimera state ($\beta = 0.1$), for the opposite effect, the injected noise annihilates the precarious and intermittent synchrony, preventing the various communities from reaching the synchrony threshold γ. Moreover, longer observational time allows for a better and more precise characterization of the Φ^R metric, confirmed by the reduced variance among the median values.

These results align with the coalition entropy plots shown in Fig. 5.19, revealing three noteworthy observations. Firstly, a more extended observational time allows for the recurrence and emergence of highly infrequent and improbable synchronous states, resulting in an increase in entropy values. Conversely, the introduction of injected noise inhibits the development of partial and intense synchrony among the networks, diminishing the occurrence of new patterns and reducing the entropy values in the case of metastable chimera state.

Finally, in contrast to the earlier scenario, the level of noise disrupts the regular patterns observed in highly synchronized states. This disruption results in the emergence of partially synchronized states in the coalition configuration, accompanied by an elevation in information uncertainty. In conclusion, the injection of noise disrupts regularity, leading to the emergence of apparent metastable chimera states or altering intermittent synchrony toward complete desynchronization.

Evidence indicates that stochastic disturbances compromise the analysis of perturbed dynamical systems when employing the measure Φ^R to evaluate dynamical complexity. This becomes especially evident in the metastable chimera state, which is marked by intermittent synchrony and has been utilized to mimic brain dynamics in various scenarios, including wakeful rest [25] and anesthesia [150]. Likewise, the precision of the integrated measure is enhanced by increasing the observational time. Short time intervals, with a limited number of samples, may not offer a comprehensive characterization of the dynamics in relation to integrated information. Despite the great importance and the theoretical background on which ΦID is established, it is essential to consider whether it is appropriate to apply such a measure to real data that exhibit behavior of interest or are usually acquired for only brief times and are known to be highly contaminated by noise, possibly dynamical. Subsequent endeavors in this direction aim to supplement the ΦID theory with a measure that considers the level of noise present in systems. This addition could assist in the accurate interpretation of the measure provided by the comprehensive integrated information measures.

5.5 Complexity vs. Dynamical Noise Power

Another crucial point that is worthwhile to emphasize is the independence between the estimated dynamical noise and the complexity of the underlying system dynamics. This fact is clearly demonstrated in Fig. 5.20, where noise estimations are conducted on 100 series, each consisting of 5000 samples obtained from two different logistic maps. The first set of series is derived from the logistic map with a parameter value of $\lambda = 3.5$, while the second set is generated by the logistic map with a parameter value of $\lambda = 4$. In both cases, a Gaussian dynamical noise with a mean of 0 and a standard deviation of 0.1 is added. Furthermore, the complexity of these time series is evaluated using the approximate entropy quantifier with an embedding dimension of $m = 2$ and a tolerance value r set to the standard value of $0.2 \times \text{std(series)}$ [132]. On the left-hand side panel, as anticipated, it is observed that noise estimations for the two groups of series show no significant differences. However, as shown on the right-hand side panel, the conventional use of ApEn fails to capture the complexity of the underlying dynamics in the presence of noise perturbations. Surprisingly, it appears that the first series is more complex than the second, which is not the case.

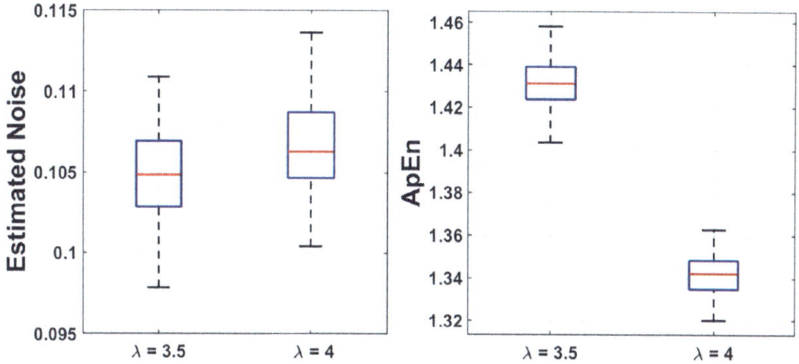

Fig. 5.20 Impact of complexity on noise estimation. The left-hand side of the figure presents boxplots illustrating absolute noise estimations for 100 logistic maps with varying complexity levels: periodic ($\lambda = 3.5$ and K-S entropy of 0) and chaotic ($\lambda = 4$ and K-S entropy greater than 0). The estimations are performed using an embedding dimension of $m = 2$ and a mean 0, Gaussian dynamical noise with a standard deviation of 0.1. Surprisingly, the estimated noises are almost indistinguishable, regardless of the complexity of the underlying system. On the right-hand side, boxplots display the complexity values of the same series using the ApEn quantifier with $m = 2$ and a radius equal to $0.2 \times$ std(series), representing a general setting

From the simulations shown in Sect. 5.5, it can be deduced that noise can impact the underlying system dynamics independently, without necessarily altering the system's complexity. Nevertheless, stochastic perturbations have the potential to introduce biases in complexity estimates, resulting in changes to the entropy of a time series. These changes can arise from alterations in the system's complexity, the underlying dynamical noise, or a combination of both. Therefore, the estimated noise remains largely unaffected by the complexity of the underlying dynamics. However, the ApEn complexity quantifier proves unreliable in capturing the true complexity of the series dynamics under noisy perturbations, leading to misleading complexity characterizations.

Hence, it is essential to consider noise estimation when assessing the complexity of a time series.

Chapter 6
Results on Real Physiological Signals

Abstract This chapter presents numerical analyses performed on real physiological systems, focusing on the characterization of their intrinsic informative noise under various pathophysiological and experimental conditions.

Each section is dedicated to the analysis of a specific physiological system, covering both cardiovascular and neural systems across different spatial scales. For each case, the real-world datasets are introduced along with a description of the experimental setup, followed by the presentation of the corresponding numerical analyses.

Physiological noise in the cardiovascular system is investigated using heart rate variability (HRV) series. The neural system is studied across multiple scales—ranging from single-neuron activity, assessed via invasive electrode recordings and calcium imaging, to cortical and whole-brain dynamics explored through noninvasive techniques such as electroencephalography (EEG) and functional magnetic resonance imaging (fMRI).

Keywords Heart rate variability · HRV · fMRI · Aging · Cam-CAN · HCP-1200 · Postural changes · Calcium imaging · Mental arithmetic · Neuronal activity

6.1 Cardiovascular Dynamics

6.1.1 Aging

6.1.1.1 The Fantasia Dataset

A total of 38 ECG signals were obtained from two groups of healthy subjects: 20 young adults aged 21–34 years old and 20 elderly adults aged 68–85 years old. The ECG signals were collected from the publicly available *Fantasia* database [71] and are accessible at https://physionet.org/content/fantasia/1.0.0/. The two groups contained equal numbers of males and females. The continuous ECG signals were recorded at a frequency of 250 Hz during 120 minutes of rest, with subjects required

A. Scarciglia et al., *Physiological Noise and Informative Randomness in Neural and Cardiovascular Systems*, https://doi.org/10.1007/978-3-031-95914-1_6

to remain awake by watching the movie *Fantasia* (Disney, 1940) [71]. The HRV series were derived from the ECG signals using the Pan–Tompkins algorithm [127] and were verified through visual inspection. The cardinality N in HRV series of the young subjects varies from $N = 5358$ to $N = 9053$ (median: $N = 7100$), while that of the elderly cohort ranges from $N = 6752$ to $N = 8410$ (median: $N = 6752$).

6.1.1.2 Experimental Setup

The aim of this experiment is to estimate the amount of physiological noise in elderly and young cardiovascular dynamics and to check if this difference is statistically significant. The noise estimation of Sect. 4.1 has been applied on the previous artifact-free HRV time series gathered from the two cohorts to retrieve the amount of stochasticity in terms of the dynamical noise standard deviation. Embedding dimensions used are $m = 2$ and $m = 3$, while the Δr tolerance step is set equal to $0.001 \times \{\text{time series range}\}$. Nonparametric Mann–Whitney tests for unpaired samples are used to compare the standard deviation noise estimates between the elderly and young groups. The null hypothesis is equal median between the two populations. The statistical comparison is performed for both embedding dimensions $m = 2$ and $m = 3$, with a significance level of $\alpha = 0.01$.

6.1.1.3 Results

The results of the noise estimation method on physiological data are presented as boxplots in Fig. 6.1 for embedding dimensions $m = 2$ and $m = 3$. The estimates of physiological noise do not exhibit significant variations with respect to the embedding dimension. A comparison of the absolute noise standard deviation achieved with the proposed method for the elderly and young groups is shown in the upper boxes of the figure. In terms of median $\pm$ median absolute deviation (MAD), the results indicate that young people have significantly higher cardiac noise levels (0.0454 ± 0.0188 for $m = 2$ and 0.0453 ± 0.0153 for $m = 3$) compared to elderly subjects (0.0219 ± 0.0058 for $m = 2$ and 0.0208 ± 0.0054 for $m = 3$), with $p < 0.01$. This difference is expected as elderly subjects tend to have a lower range in their HRV series, leading to a reduced noise level.

When the absolute noise estimation is subject-wise standardized with respect to the HRV series amplitude (as if all series have an amplitude equal to 1), the results show that young people still have higher noise standard deviation ($10.45\% \pm 2\%$ for $m = 2$ and $9.53\% \pm 1.6\%$ for $m = 3$) compared to elderly subjects ($7.2\% \pm 2\%$ for $m = 2$ and $7.15\% \pm 1.5\%$ for $m = 3$), regardless of the embedding dimension. Furthermore, noise levels in young groups exhibit more variability, as evidenced by the larger dispersion around the median value.

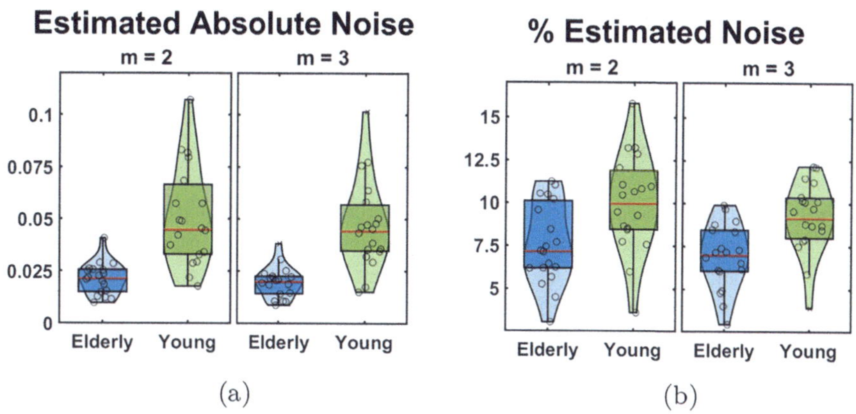

Fig. 6.1 Noise estimation in cardiovascular variability series. Boxplot statistics for the proposed physiological noise estimation in Elderly vs. Young groups: absolute noise (top figures) and standardized noise with respect to the amplitude of the series (figures below). Estimates used embedding dimension $m = 2$ and $m = 3$. Red segments stand for median values, circles denote the single-subject estimate, crosses denote the outlier value points, while "violin" plot denotes the estimate's density across central values. For both absolute and standardized noise, young cohort subjects show a higher median value with respect to the elderly ones ($p < 0.01$ for both embedding dimensions). (**a**) Absolute noise. (**b**) Standardized noise

6.1.2 Postural Changes

6.1.2.1 Dataset Description

Sample-wise noise estimation power is tested on real HRV series gathered from the *physiologic response to changes in posture* dataset [62], publicly available [50]. The dataset was generated to measure the extent of similarity in hemodynamic responses observed during stand-up, rapid, and slow head-up tilt (HUT). This initiative stems from limited understanding of the mechanisms involved in orthostatic intolerance, which refers to the failure of the hemodynamic system and reflex mechanisms to maintain blood pressure homeostasis. The dataset comprises data collected from ten healthy volunteers (five males and five females, with an average age of 28.7 ± 1.2 years). The participants gave informed written consent for the procedure, and the signals were acquired using a standard clinical ECG device. The experimental protocol consisted of a 5-minute resting state, followed by a series of postural changes, from the horizontal to the vertical position and returned to the horizontal position either through a "slow" tilt (50 s from 0 to 70°) or "fast" (i.e., 2 s from 0 to 70°), each lasting for 3 minutes. Stand-up sessions were also included. Approval for the experimental procedure was obtained from the local ethical committee, and further information can be found in [62]. The ECG data underwent analysis using the Pan–Tompkins algorithm [127] to detect the R-peaks and generate the HRV series. Additionally, the series underwent preprocessing using a point-process-based model [28] to remove ectopic and erroneous heartbeats. Further information can be found in [62].

6.1.2.2 Experimental Setup

Sample-wise nonstationary noise detection is performed with various overlapping time windows, each consisting of 150 data points. The embedding dimension is set to $m = 2$, and the radius r is adjusted to span the time series amplitude within the selected window, applying a resolution step of $\Delta r = 0.001 \times$ time series range consistently across all recordings.

The nonstationarity of the real HRV series and the nonstationarity of the related estimated noise time series, restricted to 60 s before up to 180 s after every postural change, are assessed through the Kwiatkowski–Phillips–Schmidt–Shin (KPSS) test with null hypothesis of stationarity (against the alternative hypothesis of nonstationarity) and the complementary Phillips–Perron (PP) test (null hypothesis of nonstationarity and alternative hypothesis of stationarity), with a significance threshold of 0.05.

Statistical differences between superimposed noise levels, as well as supine vs. upright positions, are assessed through the nonparametric Wilcoxon test for paired samples with the null hypothesis of equal median noise levels before and after postural changes.

6.1.2.3 Results

Nonstationarity of all the HRV and estimated noise series has been assessed by the rejection of null hypothesis for the KPSS test together with the non-rejection of the null hypothesis of the PP test. A visual representation of the nonstationarity in the estimated noise time series is shown in Fig. 6.2, where spectrograms of the estimated absolute noise time series during postural changes are reported.

Numerical results of the real data analysis are reported in Fig. 6.3 as the sample-wise median $\pm$ MAD among subjects, normalized according to the number of trials, of the encountered sample-wise noise estimation across the several trials of postural changes.

The figure illustrates the physiological noise trends during the slow, fast tilt, and stand-up transitions from the supine resting state to the vertical position. The middle panel represents the standardized noise measure with respect to the standard deviation of the signal enclosed in the window. At the bottom panel, the average RR interval $\pm$ the RR interval standard deviation across all the trials of postural changed are illustrated. In all the figures, noise estimates and HRV series have been interpolated over the recording time, with a frequency of 2 Hz, to allow comparisons among all the rest and tilt phases of the HRV recording. Starting from the postural change onset, the estimated physiological noise begins to reflect the effects of the postural changes. Descriptive and inferential statistics are reported in Table 6.1 for

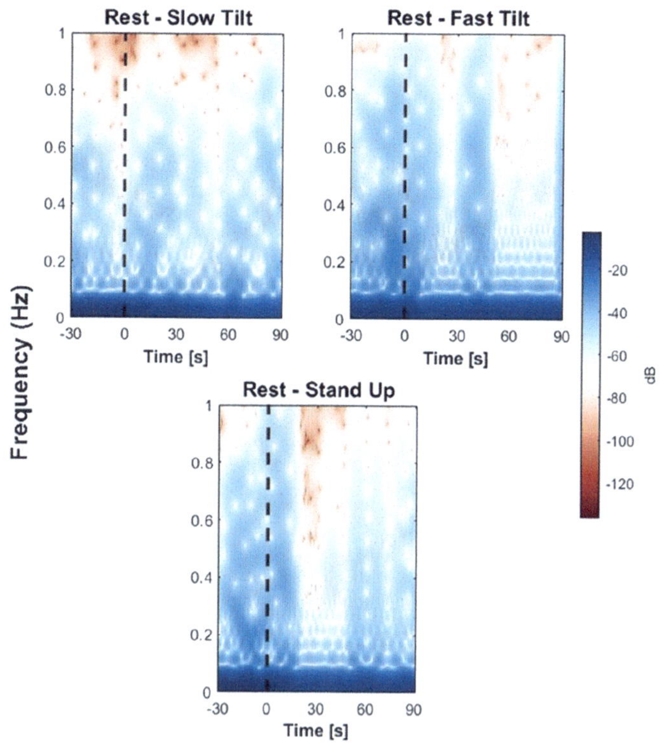

Fig. 6.2 Exemplary spectrograms of the estimated absolute noise time series. The figure represents the spectrograms of the estimated noise time series from the HRV recording of subject 12726 of the dataset of postural changes. Each panel corresponds to the first different encountered transition from rest to postural tasks. The dashed black line indicates the beginning of the task. Blue color denotes higher magnitude. Spectrograms are generated from the interpolation of absolute noise time series at 2 Hz, from 30 s before up to 90 s after the postural change from rest condition. The spectrograms are computed with the short-time Fourier transform, by using Hanning window with an amplitude of 20 s and with the temporal shift of 0.5 s—97.5% of overlap between consecutive windows

all experimental sessions. It is evident that the sliding windows capture transitions from rest to postural changes as expressed by an increase in absolute physiological noise levels, especially in case of fast and stand-up tilts. After about 60 s from the postural change onset, the absolute physiological noise levels decrease significantly (Wilcoxon test p-value less than 10^{-6}). It is possible to notice that all HRV features but the LF power are significantly affected by the postural changes.

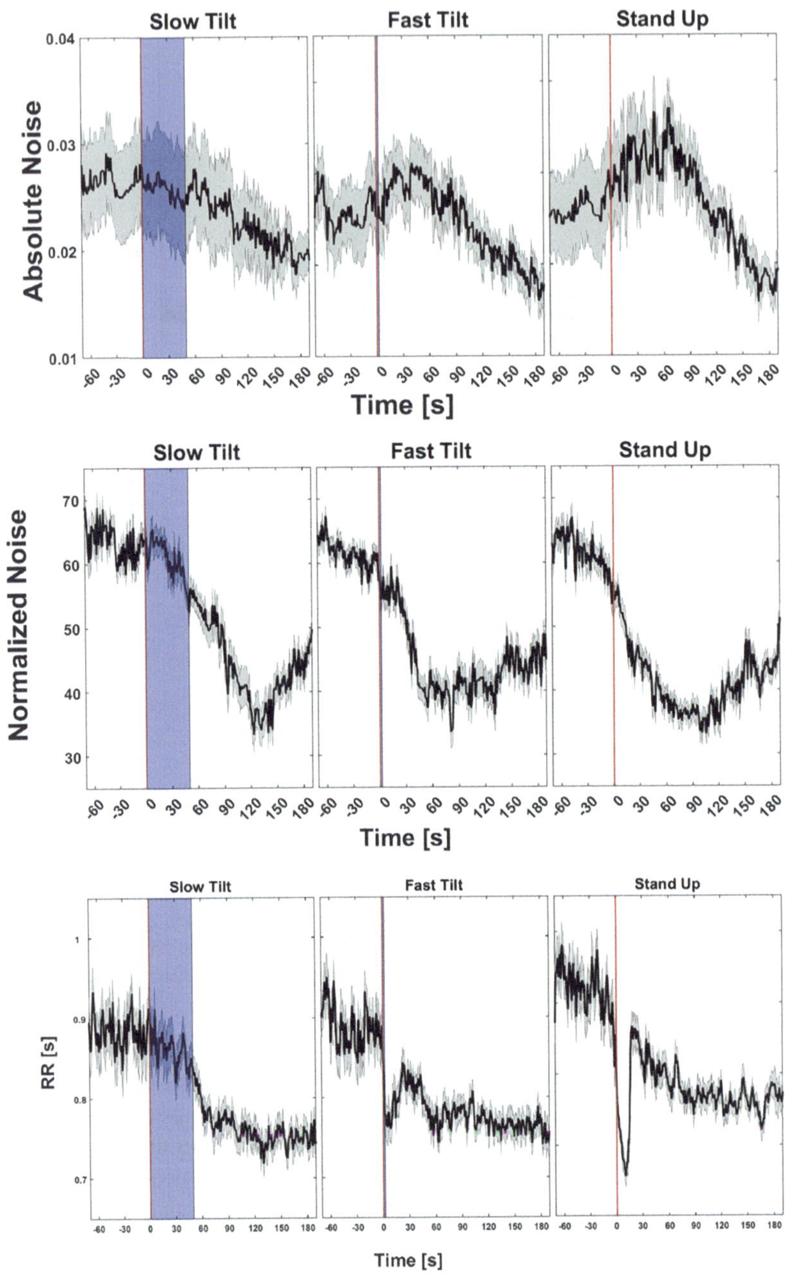

Fig. 6.3 Physiological noise in the TILT dataset. The figure shows the group-wise median (black lines) and median absolute deviation (shaded areas) values of absolute noise (upper panels) and normalized noise (middle panels), as well as RR intervals before and after the onset (indicated by the vertical red line) of postural change tasks (slow tilt, fast tilt, and stand-up). The normalized physiological noise is calculated relative to the series' standard deviation. Blue shaded areas indicate the time window during which the postural changes are performed

Table 6.1 *P*-values of resting vs. postural changes. The table reports the statistics (median and median absolute deviation) computed over a 15-second window before the task and a similar window at the end of the task, for the encountered noise level sample-wise STD, mean, RMSSD, LF and HF power. The significance of the differences in noise levels due to postural changes is confirmed by the *p*-values from the Wilcoxon tests, which test the null hypothesis of equal medians before and after the postural alterations

		Rest	Slow tilt	Rest	Fast tilt	Rest	Stand-up
Absolute noise (s)	Median	0.0264	0.0185	0.0264	0.0201	0.0273	0.0197
	MAD	0.01977	0.0084	0.0179	0.01	0.0222	0.0106
	p-Value	$<10^{-10}$		$<10^{-10}$		$<10^{-10}$	
Normalized noise	Median	62.1906	50.0448	60.6042	44.34795	59.5101	43.343
	MAD	8.7	8.23	7.34	9.84	9.24	8.56
	p-value	$<10^{-10}$		$<10^{-10}$		$<10^{-10}$	
Mean (s)	Median	0.8691	0.7529	0.905	0.7792	0.9362	0.7874
	MAD	0.13	0.0665	0.1053	0.0594	0.1106	0.068
	p-Value	$<10^{-10}$		$<10^{-10}$		$<10^{-10}$	
SDNN (s)	Median	0.0445	0.0375	0.044	0.0482	0.0451	0.0509
	MAD	0.0258	0.0146	0.0274	0.0173	0.0414	0.0174
	p-Value	$<10^{-10}$		0.1305		$<10^{-10}$	
RMSSD (s)	Median	0.0318	0.0193	0.0323	0.0205	0.0351	0.0213
	MAD	0.0309	0.0104	0.0236	0.013	0.0361	0.0131
	p-Value	$<10^{-10}$		$<10^{-10}$		$<10^{-10}$	
LF (s^2)	Median	0.001	0.0009	0.001	0.0011	0.0012	0.0016
	MAD	0.0023	0.001	0.0017	0.0018	0.0032	0.0025
	p-Value	0.3281		0.1935		0.3885	
HF (s^2)	Median	0.0005	0.0002	0.0006	0.0002	0.0006	0.0003
	MAD	0.0029	0.0007	0.0018	0.0008	0.0044	0.0013
	p-Value	$<10^{-10}$		$<10^{-10}$		$<10^{-10}$	

6.1.3 Cardiovascular Pathology

6.1.3.1 The MIT-BIH Datasets

The proposed noise estimation methodology of Sect. 4.1 and the denoising technique of Sect. 4.3 are tested on long-term heart rate variability (HRV) series from three datasets, publicly available on *physionet.org* repositories [50], and specifically the MIT-BIH database for healthy subjects (https://physionet.org/content/nsrdb/1.0.0/), Congestive Heart Failure RR datasets [89] (available at https://physionet.org/content/chf2db/1.0.0/), and MIT-BIH Atrial Fibrillation database [1] (and https://physionet.org/content/afdb/1.0.0/) [50, 118].

More in detail, the first database is comprised of 18 healthy subjects (age range 20–50 yo, with ECG sampled at 128 Hz) with normal sinus rhythms (NS data), the second dataset counted 29 HRV series gathered from subjects with congestive heart failure (CHF data) (24-hour Holter recordings sampled at 128 Hz, 34–79 yo), and the third dataset is comprised of 23 HRV long-term (i.e., 10-hour ECG series sampled at 250 Hz) series gathered from subjects with atrial fibrillation (AF dataset). Further details on experimental recordings are available on the relative web pages.

HRV series are derived from ECG signals through the well-known Pan–Tompkins algorithm for the identification of R-peaks [127] and visually inspected for artifacts which were eventually corrected through a point-process-based method [28].

6.1.3.2 Experimental Setup 1: Physiological Noise Detection

The monovariate noise detection algorithm is applied on all the HRV recordings gathered from the three datasets above.

To enhance the resolution of the HRV series in the CHF dataset, an interpolation of the HRV time series into a finer grid with 10 time samples between each heartbeat and its consecutive one is performed, forming new series by taking successive interpolated values.

To avoid any potential bias associated with the series length, an artifact-free 10,000-sample segment is selected for each subject. Previous studies [32] have shown that the three cohorts exhibit different complexity levels as measured by an entropy measure over multiple scales, specifically, the multiscale sample entropy, which measures the regularity of the series at various time lags. The results indicate that for time scale one, the complexity measures for the healthy cohort were smaller than those for the remaining two cohorts. However, this trend is inverted for higher time scale factors, indicating a change in complexity. This change in complexity is also supported by other quantifiers, such as symbolic entropy [7].

The noise estimation algorithm has been applied with an embedding dimension $m = 2$ and a $\Delta r = 0.001 \times \mathcal{R}$.

Significant differences for group-wise statistics are established through non-parametric Kruskal–Wallis test with threshold $\alpha = 0.05$, with null hypothesis of equal median among all the cohorts; post hoc pairwise comparisons are performed, instead, through nonparametric Mann–Whitney test, with threshold $\alpha = 0.05/3 \approx 0.01$ and null hypothesis of having equal median between two groups. Moreover, in order to test the robustness of the method, a conservative surrogate data analysis with null hypothesis of not having physiological noise is employed. In this context, for any time series, 50 surrogate data series are generated using random permutations, where the system dynamics is destroyed. The null hypothesis is rejected when the noise level for the original series is before the 2.5-th and after the 97.5-th percentiles of the noise distribution for its surrogate data. Of note, given that noise levels typically increase in series with destroyed dynamics, as one would intuitively expect, employing a one-tail surrogate data approach can also effectively serve as a means to assess the method's robustness.

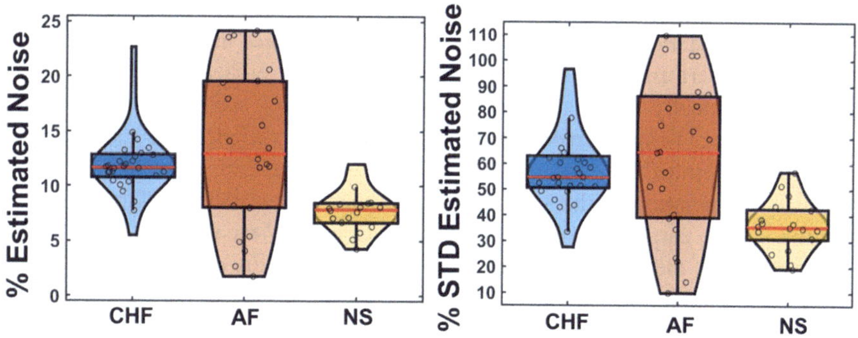

Fig. 6.4 Noise estimation in cardiovascular variability series. Boxplots statistics for the proposed physiological noise estimation in CHF, AF, and NS groups with embedding dimension $m = 2$. In the left panel, noise estimates are standardized with respect to the signal range, whereas in the right panel, noise estimates are normalized by the series standard deviation. Red horizontal lines indicate group median values, whereas circles indicate the single-subject estimate, and the violin plot denotes the estimate's density across central values

6.1.3.3 Results

Noise estimates from cardiovascular activity series are summarized in Fig. 6.4. Cardiovascular pathologies modulate underlying physiological noise, particularly increasing in AF group (13.56% ± 5.98%) and CHF group (11.67% ± 1.93%), with respect to healthy NS group (7.85% ± 1.39%) heartbeat dynamics. Moreover, cardiovascular noise reaches about 50% of HRV series standard deviation in case of AF (64.5% ± 24.9%) and CHF (54.74% ± 10.69% with respect to healthy subjects (35.5% ± 7.18%).

Interestingly, groups differ in median values and in dispersion, with NS group being close to the median value, and the other two having higher intersubject variability, more accentuated in AF group. Group-wise statistics confirm significant differences through nonparametric Kruskal–Wallis test (p-value < 0.00298), whereas post hoc comparisons performed through nonparametric Mann–Whitney test confirmed main differences between CHF and NS groups (p-value < 0.0095) and between AF and NS groups (p-value < 0.0031). Furthermore, the application of surrogate data analysis has effectively verified noise as a discriminant statistic for the entirety of the HRV series. This conclusion is drawn from the rejection of the null hypothesis stating that the original and surrogate series possess equal noise levels.

6.1.3.4 Experimental Setup 2: Denoising

Each HRV time series is assumed to be generated by a dynamical system subject to physiological noise, reflecting interactions with other physiological systems. The entire series are segmented into non-overlapping windows of 5000 samples each to mitigate nonstationarity and physiological state changes in lengthy recordings. Denoising is performed within each window using an embedding dimension of $m = 2$. Additionally, various metrics are computed within each window, including Sample Entropy [137], estimated on non-overlapping sub-windows of 1000 samples each, with an embedding dimension of $m = 2$ and a radius equal to 0.15 times the standard deviation of the series. This approach enhances the reliability of the SampEn metric, particularly effective for short-term HRV recordings [137].

The choice of the sample entropy as nonlinear complexity quantifier relies on its primary applicability on HRV series. The aim is to establish how much the denoising procedure can regulate the complexity of the cardiovascular dynamics and to see how the complexity behaves when noise is reduced for different pathophysiological conditions. Moreover, the power spectrum for the low frequency (LF) and high frequency (HF) bands (0.04–0.15 Hz and 0.15–0.4 Hz for LF and HF, respectively) is computed via the FFT algorithm on the HRV interpolated series at 4 Hz. Furthermore, noise levels were estimated before and after noise reduction using the method proposed in Sect. 4.3. Additionally, the mean and standard deviation of the signals are computed. For each cardiovascular series, only one value from the window estimations is considered by averaging across the non-overlapping windows.

6.1.3.5 Results

The performance of the physiological denoising technique on heartbeat dynamics series has been tested through the Sample Entropy and physiological noise estimation $\bar{\sigma}$ (Fig. 6.5) quantifiers. In addition, standard HRV features defined in the time

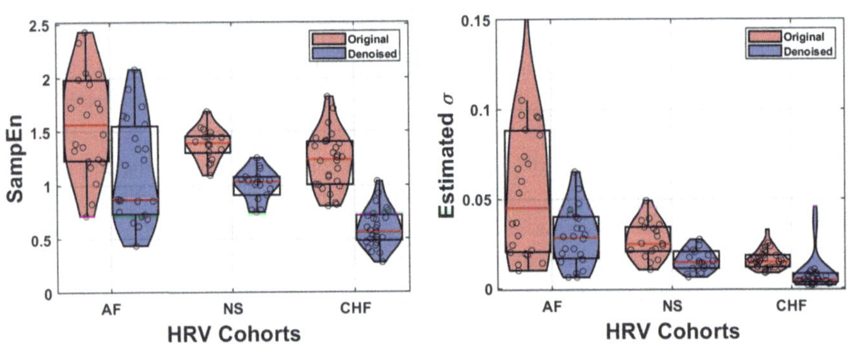

Fig. 6.5 Noise reduction in HRV data. The two panels show the box and violin plots, used for comparing probability distributions, for the Sample Entropy (top) and the estimated noise level in the series (bottom). The three pathophysiological conditions are atrial fibrillation (AF), health state (NS), and congestive heart failure (CHF) for the original and denoised series (red and blue violin plots, respectively). The values reported here are gathered by taking the average of the two quantifiers computed for non-overlapped windows of 5000 samples for each HRV series

domain, such as the statistical mean and variance and frequency domain features including the power within the LF and HF bands (see Table 6.2) were evaluated.

Figure 6.5 depicts a median Sample Entropy value drop after noise reduction with respect to the original signals for the three cohorts: patients with atrial fibrillation (AF), healthy (NS), and subjects who experienced congestive heart failure (CHF); the same trend has been seen for the noise standard deviation $\bar{\sigma}$ estimated through the method in Sect. 4.1. Interestingly, the median values for the AF and NS groups are inverted after denoising. The values lowering is also significantly observed for the LF and HF bands' power—$p < 0.05/6$ from Wilcoxon test with Bonferroni correction—as depicted in Table 6.2. Only the statistical mean appears to be unaffected by the noise reduction process ($p > 0.05/6$), whereas the reduction in variance is not large but remains significant.

Exemplary denoised HRV series are depicted with blue lines in Fig. 6.6. HRV series of the atrial fibrillation dataset are more smoothed because higher noise levels are associated with these series (see Table 6.2 and [148]).

6.2 Cortical Dynamics

6.2.1 Mental Arithmetic

The monovariate noise estimation methodology of Sect. 4.1 is further tested on a publicly available EEG series dataset, namely the *EEG During Mental Arithmetic Tasks* [198]. This dataset is comprised of EEG gatherings from 36 healthy volunteers undergoing a 180 s resting phase and a 60 s mental cognitive workload task (i.e., performing mental arithmetic). Recordings from four subjects were rejected after visual inspection owing to gross artifacts. Thus, data from 32 persons (24 females) with ages of 18 ± 2.01 yrs on average were retained for further processing. The electrophysiological signals were sampled at 500 Hz. The eligibility criteria were normal or corrected-to-normal visual acuity, normal color vision, no clinical manifestations of mental or cognitive impairment, and no learning disabilities. The use of psychoactive medication, drug or alcohol addiction, and psychiatric or neurological complaints were considered as additional exclusion criteria. Power line notch (50 Hz) and [0.5 Hz–45 Hz] band-pass filters were applied in the EEG series before independent component analysis and were used to identify artifacts (i.e., eyes, muscles, and cardiac pulsation) that were subsequently rejected. Further details on signal acquisition and preprocessing can be found in [198]. For each EEG series, outlier samples over the 98th percentile are not retained for further processing. To avoid bias associated with the series length, it is important to note that the length of the time series can affect the accuracy of noise estimation [149].

Table 6.2 Comprehensive results of the HRV noise reduction. The table reports the statistics for the LF and HF power, statistical mean and variance (VAR), sample entropy, and noise estimation $\bar{\sigma}$ for the HRV datasets, before and after the noise reduction process, expressed in the appropriate units. For each indicator, we report the median value $\pm$ MAD along with the range. Again, the values reported here are gathered by taking the average of the two quantifiers computed for non-overlapped windows of 5000 samples for each HRV series. Bold values indicate significant changes between original vs. clear series per group with $p < 0.05/6$ from Wilcoxon test

		AF		NS		CHF	
		Original	Clean	Original	Clean	Original	Clean
LF [ms^2]	Median ± MAD	**1618.09 ± 3112.63**	**1292.29 ± 1031.27**	**1854.57 ± 1113.17**	**1546.40 ± 781.68**	**227.21 ± 678.66**	**189.07 ± 605.38**
	Range	**[415.62, 24702.84]**	**[185.96, 6329.31]**	**[1107.67, 6083.10]**	**[865.13, 4099.09]**	**[23.35, 4278.27]**	**[16.52, 3915.29]**
HF [ms^2]	Median ± MAD	**2850.65 ± 4542.62**	**1919.16 ± 921.34**	**1038.11 ± 1056.45**	**509.07 ± 625.32**	**357.74 ± 1034.75**	**242.16 ± 969.79**
	Range	**[663.78, 35653.30]**	**[386.63, 5334.64]**	**[106.95, 6057.79]**	**[82.37, 4522.96]**	**[60.42, 9568.77]**	**[47.07, 9015.82]**
Mean [ms]	Median ± MAD	744.82 ± 123.88	743.57 ± 124.14	785.68 ± 71.57	785.82 ± 71.57	689.60 ± 73.06	689.55 ± 73.03
	Range	[577.86, 993.33]	[577.82, 993.44]	[688.19, 957.84]	[688.15, 957.78]	[549.87, 872.39]	[549.92, 872.46]
Variance [ms^2]	Median ± MAD	**10360.89 ± 6476.13**	**5392.34 ± 2938.66**	**4909.25 ± 3468.27**	**4162.74 ± 2780.54**	**2235.26 ± 2925.46**	**1880.56 ± 2756.73**
	Range	**[1247.51, 45832.93]**	**[579.72, 18188.26]**	**[2330.62, 13615.67]**	**[2138.59, 10523.02]**	**[240.11, 21590.60]**	**[154.74, 20782.22]**
SampEnt	Median ± MAD	**1.66 ± 0.39**	**0.88 ± 0.41**	**1.40 ± 0.11**	**1.04 ± 0.11**	**1.27 ± 0.21**	**0.59 ± 0.14**
	Range	**[0.71, 2.43]**	**[0.44, 2.08]**	**[1.09, 1.68]**	**[0.75, 1.25]**	**[0.80, 1.82]**	**[0.28, 1.04]**
$\bar{\sigma}$ [ms]	Median ± MAD	**53.47 ± 34.76**	**28.95 ± 12.50**	**24.78 ± 8.23**	**14.60 ± 4.72**	**15.02 ± 4.10**	**4.87 ± 6.45**
	Range	**[10.03, 200.84]**	**[6.20, 65.07]**	**[10.34, 48.97]**	**[6.19, 27.10]**	**[8.50, 32.50]**	**[1.56, 45.38]**

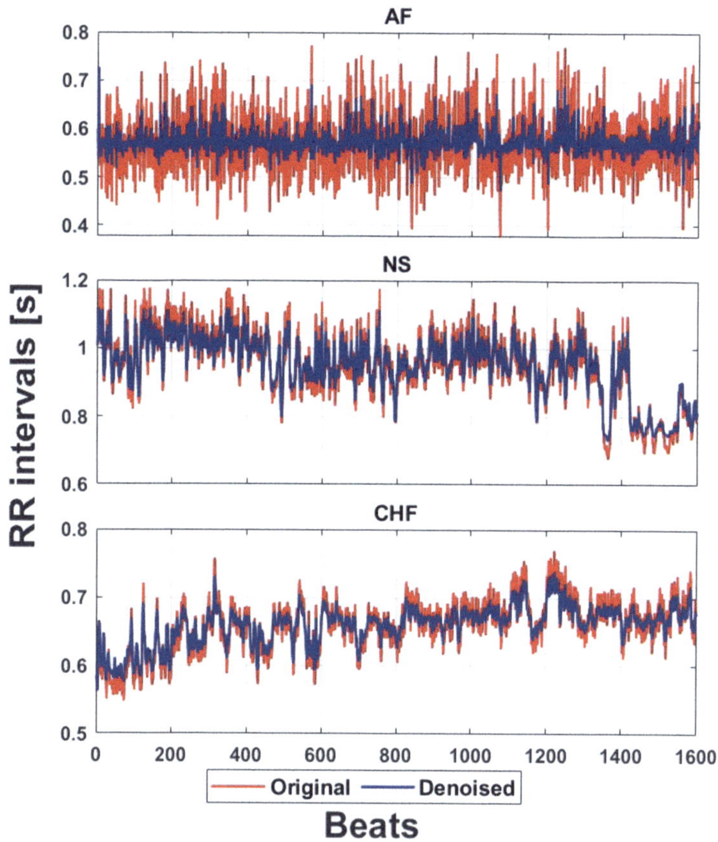

Fig. 6.6 Exemplary noise reduction in HRV series. The three panels show the noise reduction in three exemplary HRV series, gathered, respectively, from the AF, NS, and CHF cohorts. The original series is depicted in red, while the denoised is in blue. Each series comprises 1600 beats (x-axis) with interbeat time interval reported on the y-axis

6.2.1.1 Experimental Setup

Therefore, the focus is on the central minute of the 180-second resting state recordings to minimize the impact of any potential noise caused by transitions to and from rest. For each channel and task, the EEG series contained 30,000 samples. For any of the 19 scalp channels, Wilcoxon statistical test ($\alpha = 0.05$), with null hypothesis of having equal median between the noise levels computed in resting and computational tasks, is used to assert the significance for different levels of noise together with the Bonferroni–Holm correction for p-values [65]. The robustness of the method has also been tested with a surrogate data approach, following the same procedure as the HRV dataset. To estimate the noise level in the EEG data, for every

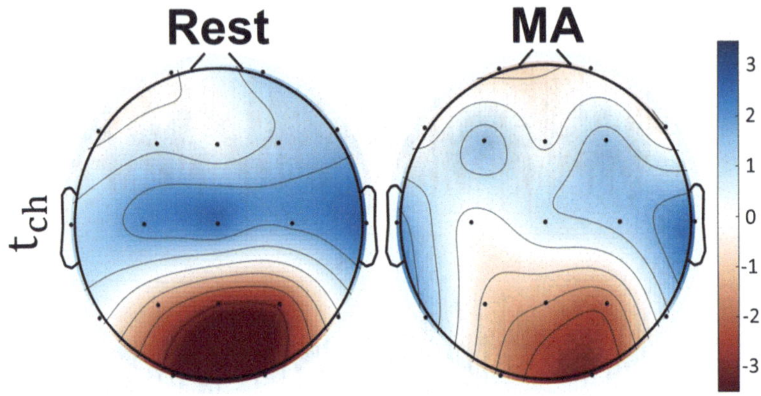

Fig. 6.7 Topographical representation of EEG noise distribution during the resting state and mental arithmetic task. Normalized median values across subjects as t-scores across channels using an embedding dimension $m = 5$

channel and experimental protocol, an embedding dimension of $m = 5$ and a time delay of $\Delta r = 0.001$ times the range of the time series are used.

Previous studies showed that higher gamma activity is associated with mental stress tasks, while alpha band was mainly present at resting state; from a time-domain analysis, it has been seen that fractal measures are capable of better differentiating signals between rest and task phases [184].

6.2.1.2 Results

Experimental results on noise estimates from EEG series are here reported. Firstly, Fig. 6.7 reports a topographical representation of noise estimates across the scalp during resting and mental arithmetic states. More specifically, the represented values refer to a t-score normalization performed across channels to see if significant variation were present across different scalp locations. For each channel, the value has been averaged across subjects. From this representation it is possible to appreciate that occipital and posterior parietal channels have the highest cortical noise, compared to the central electrodes, which have the lowest. In addition, right hemisphere seems to have extreme variation, with both the highest value on the posterior scalp and the lowest on the temporal lobe.

Variations between experimental phases are also represented in Fig. 6.8. In particular, Fig. 6.8a, b illustrates a topographical representation of median (top panels) and median absolute deviation (bottom panels), calculated across subjects, for both experimental conditions, i.e., resting state (left column) and mental arithmetic (right column), with embedding dimension $m = 5$. Figure 6.8c shows results from the statistical comparison performed through nonparametric Wilcoxon test for paired samples, where p-values were corrected for multiple comparisons in accordance

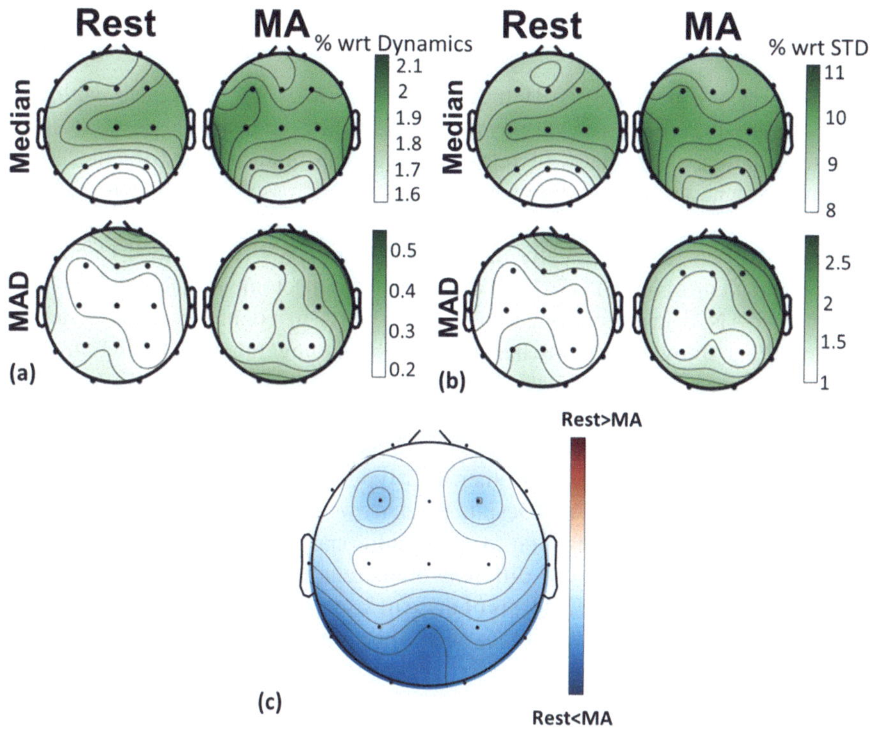

Fig. 6.8 Topographic representation of EEG noise estimations and statistical comparison. (**a**)–(**b**) Median (top row) and median absolute deviation (MAD) (bottom row) cortical noise levels across subjects during the resting state (left column) and the mental arithmetic (MA) task (right column). The color bar in panel (**a**) represents the standardized noise estimations expressed as percentage with respect to the range of the series. In panel (**b**) the color bar represents the percentage of estimated noise normalized with respect to the standard deviation of the series. (**c**) The topographical representation shows the corrected p-values obtained from the nonparametric Wilcoxon test for paired samples. White areas indicate nonsignificant differences, and blue color indicates higher noise during the MA task. The statistical significance is set at $\alpha = 0.05$, and darker areas represent lower p-values

with Bonferroni–Holm criterion [65]. Mental arithmetic task is associated with higher neurophysiological noise levels with respect to the resting state, especially in the prefrontal and occipital regions. Cortical noise comprises about 1.5–2% of EEG dynamics (about 8–11% of signal standard deviation) group-wise. Tasks differ in median values and in dispersion, with mental arithmetic associated with higher intersubject variability over the frontal, right temporal, and occipital regions. The surrogate data approach provides compelling evidence that physiological noise can be used as a discriminating statistic for all EEG recordings. This is achieved by rejecting the null hypothesis that the original and surrogate series have equal noise levels.

6.3 Brain Dynamics

Here it is assumed that in MRI measurement noise is not associated with age and explore whether age-related physiological changes in the brain are associated with modulation of estimates of intrinsic dynamical brain noise. To achieve this objective, stochastic brain components are analyzed across different stages of brain maturation, leveraging high-quality publicly available fMRI datasets. Two publicly available datasets are specifically utilized: the Cam-CAN (Cambridge Centre for Ageing and Neuroscience) [174] and the Human Connectome Project (HCP) [191]. The Cam-CAN dataset, designed to explore shifts in cognitive abilities during healthy aging, provides comprehensive data, including functional MRI, from a population-based cohort of over 700 healthy individuals aged 18–87 years. In contrast, the HCP dataset aims to provide a comprehensive understanding of the structural and functional connectivity of the human brain. It includes high-resolution functional MRI data, along with other data types, from more than 1200 healthy young adults aged 22–35 years. Despite the narrower age range in the HCP dataset, it is selected for this investigation to offer additional insights into the modulation of intrinsic brain noise across different brain regions and age groups within a substantial data sample. The analysis of these datasets is intended to provide insights into the contribution of intrinsic brain noise to healthy aging and brain function.

6.3.1 fMRI Cam-CAN Dataset

The dataset consists of multimodal fMRI series collected from a large (approximately $N = 700$), cross-sectional adult lifespan (18–87 years old) population-based sample. The aim here is to characterize age-related changes in cognition and brain structure and function and to uncover the neurocognitive mechanisms that support healthy cognitive aging. For each subject, fMRI recordings in three functional runs (resting state with a repetition time, TR, of 1970 ms, movie-watching with a TR of 2470 ms, and a sensorimotor task with a TR of 1970 ms) were collected. In a subsequent stage, fMRI data were preprocessed and regrouped into 116 ROI series according to the Automated Anatomical Labeling (AAL)1 atlas [45].

Here, data from 645 subjects who have completed fMRI series for resting state task are included. Each fMRI series contains $N = 216$ samples for each ROI, for every subject. Therefore, for each subject, intrinsic brain noise is estimated in 116 ROI-wise fMRI series. Further details on the data collection and characteristics can be found in [174].

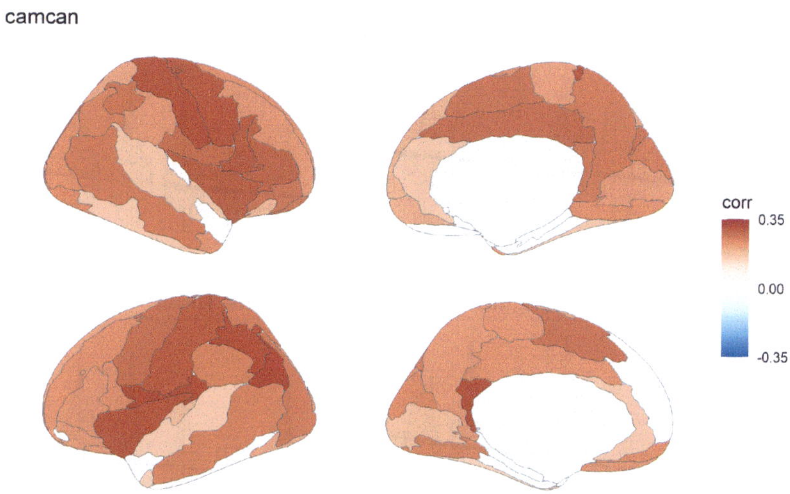

Fig. 6.9 Brain ROIs with a significant correlation between noise and age. The brain images depict the first 78 ROIs (excluding the limbic system, cerebellum, and vermis), illustrating the significant Spearman correlation coefficients between noise and aging. These images correspond to the Cam-CAN dataset, where darker regions indicate stronger correlations

6.3.1.1 Cam-CAN Dataset: Results

For the Cam-CAN dataset, the Spearman correlation coefficient is calculated between the estimated noise power-to-signal ratio and subjects' age for each ROI. Intrinsic brain noise power in these fMRI series reaches up to the power of the signal and varies among brain areas and age. To illustrate, in Fig. 6.9 the ROIs showing a significant correlation between noise and age are represented. From the dark red areas, it is evident that the correlations are mostly verified in the precentral and frontal areas. More details about the Spearman correlation coefficient and associated p-value for any ROI between subject-wise noise estimates and subjects' ages, alongside the AAL1 reference abbreviation and corresponding MNI coordinates, are provided in [147].

A total of 73 ROIs exhibit statistically significant positive correlations between noise and age, while only two ROIs (# 71 and # 72) show a significant, negative trend. The left-hand side column of Fig. 6.10 displays exemplary ROI statistics (#12 = Frontal Inferior Operculum R, #14 = Triangular Inferior Frontal Gyrus R, and #16 = Frontal Inferior Orbit R) with statistically significant Spearman correlation coefficients ($\rho = 0.2843, 0.2763$, and 0.2573, respectively) and $p <$ $(0.05/116)$ computed subject-wise, following a Bonferroni correction. It is noteworthy that intrinsic brain noise can contribute substantially to the overall signal power, ranging from 10% to 100% of the fMRI signal power. In addition, the estimated brain noise level seems to positively correlate with the fMRI signal power for any ROI, with a mean $\pm$ standard deviation correlation coefficient of 0.66 ± 0.078 among

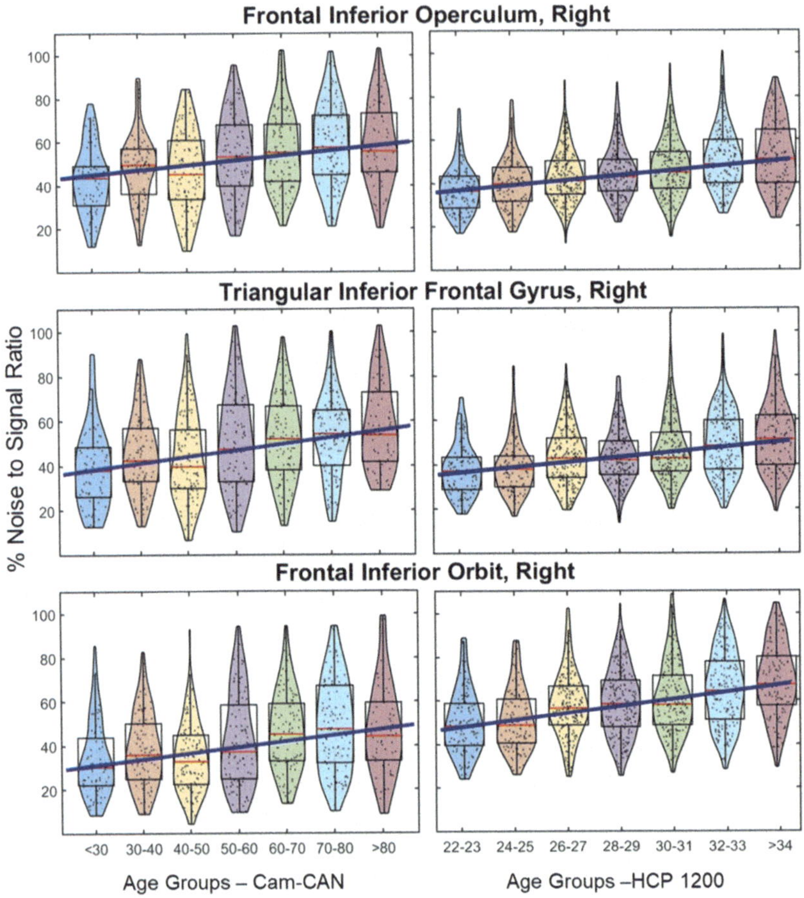

Fig. 6.10 Age-related fMRI noise variation in key ROIs within the Cam-CAN and HCP 1200 datasets. The figure presents boxplot statistics of the noise-to-signal ratio, computed as the variance of the encountered noise divided by the power of the signal, for three significant regions of interest (ROIs): Frontal Inferior Operculum R (#12), Triangular Inferior Frontal Gyrus R (#14), and Frontal Inferior Orbit R (#16). The left-hand column corresponds to the Cam-CAN dataset, while the right-hand column represents the first recording of the four HCP-1200 trials. In the left-hand side panels, the figure shows the variations in noise-to-signal ratio across different age groups (18–87) for the Cam-CAN dataset. The age groups are categorized as follows: <30 (73 subjects), 30–40 (95 subjects), 40–50 (110 subjects), 50–60 (95 subjects), 60–70 (104 subjects), 70–80 (107 subjects), and >80 (61 subjects). Individual subject noise estimations are denoted by black crosses. All the ROIs exhibit a significant positive trend in median noise levels as age increases, with Spearman coefficients of $\rho = 0.9286, 0.9286$, and 0.87 ($p = 0.0067$, 0.0067, and 0.0238, respectively). Similarly, the right-hand side panels illustrate the relationship between the noise-to-signal ratio and age range series for the first recording of the HCP-1200 trials. The age groups are divided as follows: 22–23 (114 subjects), 24–25 (113 subjects), 26–27 (195 subjects), 28–29 (185 subjects), 30–31 (177 subjects), 32–33 (177 subjects), and >34 (121 subjects). In this case as well, the boxplots demonstrate significant positive trends in median noise levels as age increases, with Spearman coefficients of $\rho = 1.000, 0.8929$, and 0.9643 ($p < 0.0123$). Across all panels, the thick blue line represents the least-square linear regression of the group-wise median noise levels

all the ROIs, indicating that the brain noise significantly contributes the fMRI signal variability. The detailed results can be found in [147].

6.3.2 fMRI HCP1200 Dataset: Regions of Interest

Publicly available data are used from the Human Connectome Project (HCP), which includes a total of 1086 subjects. The data can be accessed at https://www.humanconnectomeproject.org. In this analysis, the focus is on the 116 fMRI node time series that correspond to the 116 ROI, as defined by the AAL1 atlas [45]—consistently with the Cam-CAN setup. The HCP dataset consists of fMRI recordings obtained from two resting state sessions (Rest1 and Rest2). Each session involved scanning from left to right (LR) and from right to left (RL), resulting in a total of four fMRI recordings for each subject. Each ROI fMRI time series comprised 1200 volumes, with a repetition time (TR) of 720 ms. This equates to approximately 15 minutes of scanning for each of the four experimental settings.

6.3.2.1 fMRI HCP1200 Dataset Results

Noise-to-signal ratio estimations have been conducted for all four recordings of the HCP dataset across different tasks. Additionally, the Spearman correlation coefficient has been computed between the estimated noise power and the subjects' age for each ROI in all sessions. A matching analysis is conducted to compare results from the four different resting state sessions. A match for an ROI is determined if the noise levels of two different datasets either significantly changed with age (Spearman $p < 0.05/116$) or did not exhibit a joint correlation with age (Spearman $p > 0.05/116$). The results in Table 6.3 reveal that there were no significant differences in noise levels across the various recordings, with at least 90% of the ROIs displaying matching trends between different resting state sessions.

Therefore, for the sake of brevity and without loss of generality, further details and analyses are provided on the HCP dataset focusing on the first recording in the resting state, denoted as "Rest1," which is scanned from left to right (LR). Specifically, in Fig. 6.11 it is highlighted the Spearman correlation coefficient

Table 6.3 Percentages of ROIs with consistent correlation among the different recordings of HCP-1200

Percentages of matches	HCP_REST1_LR	HCP_REST1_RL	HCP_REST2_LR	HCP_REST2_RL
HCP_REST1_LR	100.00%	89.66%	93.10%	93.10%
HCP_REST1_RL	89.66%	100.00%	89.66%	86.21%
HCP_REST2_LR	93.10%	89.66%	100.00%	91.38%
HCP_REST2_RL	93.10%	86.21%	91.38%	100.00%

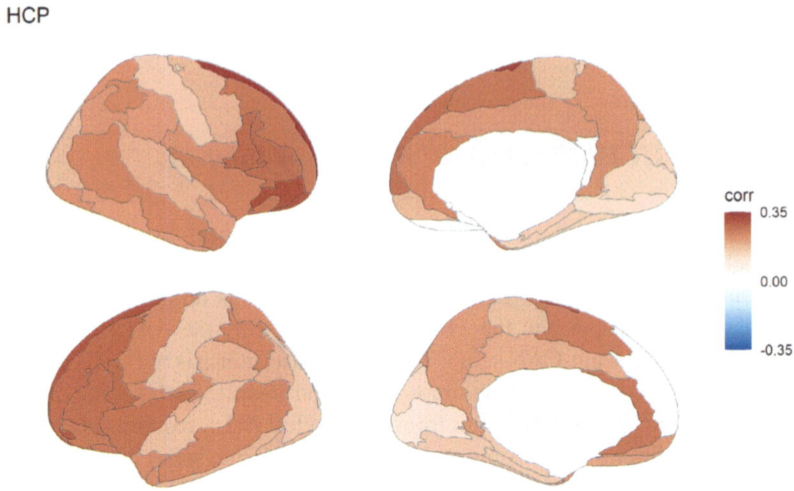

Fig. 6.11 Brain ROIs with a significant correlation between noise and age. The brain images represent the first 78 ROIs—limbic system, cerebellum, and vermis are excluded—where the significant Spearman correlation coefficient is represented between noise and aging. The upper four brain images are associated with the Cam-CAN dataset, the remaining ones with the HCP-1200 dataset. The darker the areas, the higher are the correlations

between the estimated noise power and the subjects' age for each ROI in this particular recording.

Among the 116 ROIs analyzed, 92 of them show statistically significant positive correlations between noise and age. As for the Cam-CAN dataset, Fig. 6.10 illustrates this relationship, specifically highlighting three ROIs: Frontal Inferior Operculum R (#12), Triangular Inferior Frontal Gyrus R (#14), and Frontal Inferior Orbit R (#16). These ROIs exhibit statistically significant Spearman correlation coefficients ($\rho = 0.3276, 0.3027$, and 0.328, respectively) with corresponding subject-wise p-values of less than $(0.05/116)$, computed following a Bonferroni correction.

For a comprehensive overview of the Spearman correlation coefficients and associated p-values between subject-wise noise estimates and subjects' ages, please refer to [147]. This table provides the results for all ROIs, along with the AAL1 reference abbreviation and corresponding MNI coordinates. Similarly to the previous analysis, positive correlations are present between estimated brain noise levels and fMRI signal variance, with a mean and standard deviation of 0.85 ± 0.074. This confirms that noise plays a fundamental role in the series variability. The results can be found in [147].

6.3.3 *fMRI HCP1200 Dataset: Nodes*

Additional analysis is conducted on the HCP dataset, which involved 25 fMRI node time series obtained from each subject. These node time series corresponded to the 25 components resulting from the group-ICA decomposition. Each node time series consisted of $N = 1200$ samples from four resting state fMRI runs. The HCP consortium performed preprocessing on the 15-minute resting state fMRI data of each subject, employing a combination of FSL Multivariate Exploratory Linear Optimized Decomposition into Independent Components (MELODIC) (ICA) and FMRIB's ICA-based X-noiseifier (FIX) to remove artifacts. Intersubject registration of the cerebral cortex was accomplished using areal-feature-based alignment and the Multimodal Surface Matching algorithm. For the input of group principal component analysis (PCA), each dataset underwent temporal demeaning and variance normalization. Specifically, the HCP consortium utilized Multivariate Imputation of Gaussian Process (MIGP) group-PCA output as input for FSL's MELODIC tool for group-ICA, applying spatial ICA at various dimensionalities. For the analysis, dimensionality is 25, which determined the number of distinct ICA components retained for further investigation. Spatial ICA was conducted in grayordinate space, encompassing surface vertices and subcortical gray matter voxels. The set of ICA spatial maps was then projected onto each subject's fMRI time series data to derive one representative time series per ICA component (referred to as a network "node" here). The noise level is estimated for each of the 25 components in each run and then retained the average across the four runs for subsequent statistical analysis.

6.3.3.1 fMRI HCP1200 Dataset: Nodes Results

Noise estimation is further performed on fMRI series obtained from 25 highly aggregated nodes. For the purpose of brevity and due to the similarity of the noise estimation across the four recordings in the resting state of the HCP-1200 Dataset, the results are presented specifically for the recording in task "Rest1," scanned from LR. The findings are summarized through group-wise boxplot statistics, as depicted in Fig. 6.12. The significance level for the test is set at $\alpha = 0.05/25$, considering the Bonferroni correction.

From a spatial perspective, intrinsic brain noise contributes to the entirety of fMRI variance and demonstrates variability across different brain regions or group-ICA clusters. On average, noise levels range from 20% to 40% for most nodes, except for clusters 21, 23, and 24, which exhibit median noise power-to-signal power ratios ranging from 50% to 90%.

From the perspective of aging, the fMRI series and the associated stochastic brain dynamics are classified into three age ranges: 22–25, 26–30, and >30. Across all nodes, a consistent pattern of increasing noise-to-signal ratios with aging is present. This increasing trend is confirmed by conducting a Kruskal–Wallis test, where the null hypothesis assumed equal medians among the three age cohorts.

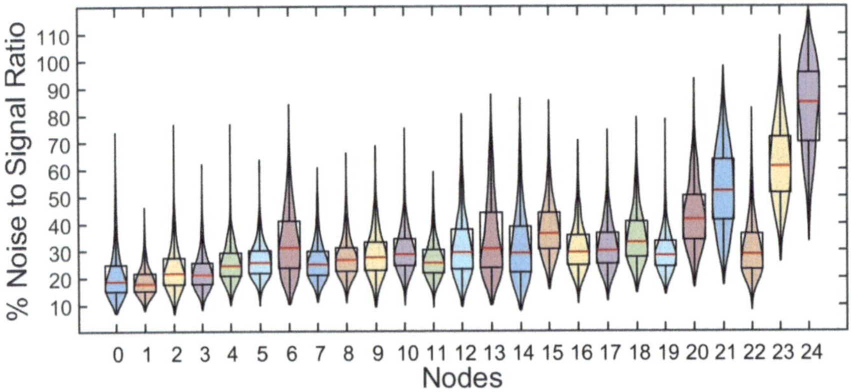

Fig. 6.12 fMRI time series physiological noise analysis. Displayed are the boxplot statistics for each group-ICA component within the HCP1200 dataset, considering an embedding dimension of $m = 2$

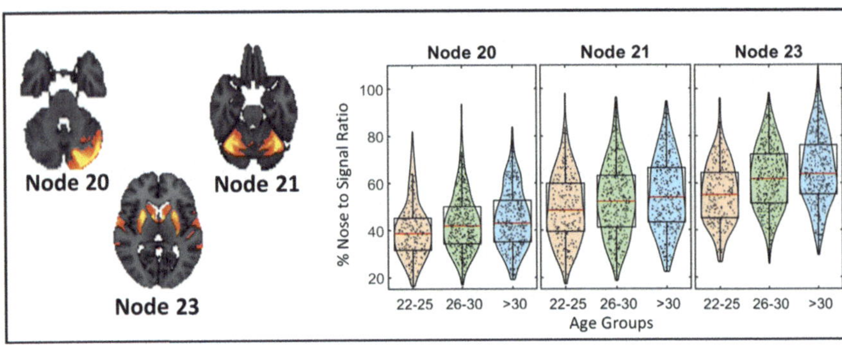

Fig. 6.13 Age-related noise variation in fMRI cluster nodes. This figure illustrates exemplary node-wise noise trends in relation to aging, using an embedding dimension of $m = 2$. Black crosses represent noise estimation for individual subjects, while the horizontal red line signifies the median noise level per age range. Age groups are divided into 22–25 (218 subjects), 25–30 (432 subjects), and >30 (353 subjects)

Exemplary boxplot statistics for nodes exhibiting higher brain noise and displaying significant noise trends with age are presented in Fig. 6.13. For more comprehensive information, please refer to [147].

6.4 Neuronal Dynamics

After demonstrating the reliability of the monovariate estimation method using synthetic data from Izhikevich's neural models and simulated oscillatory calcium dynamics driven by dynamic noise, the proposed method of Sect. 4.1 is applied on experimental time series obtained from calcium imaging in diverse brain regions of mice and zebrafish and to electrophysiological data from a multisite 128-channel cortical probe in anesthetized rats.

6.4.1 Neural Spiking Activity Based on Calcium Imaging Dataset

Calcium imaging is a microscopy technique used to optically measure the intracellular calcium shift operated by different cellular mechanisms. It is widely used as a proxy of neural activity. For monitoring calcium signaling, several procedures have been developed: Neurons can be labeled with chemical indicators or protein-based genetically encoded calcium indicators which emit fluorescence in response to changes in calcium concentration [93]. Recently, a ground-truth database of calcium imaging recordings has been published by Rupprecht et al., along with a method for directly quantifying the noise from the raw fluorescence-based signal constituting the datasets [140]. Data can be found at https://github.com/HelmchenLabSoftware/Cascade/tree/master. Using Rupprecht et al.'s datasets, the proposed noise estimation method has been compared and complemented with their fluorescence-based one. They use fluorescence signal F to estimate the noise level as follows [140]:

$$\nu = \text{Median}_t |(F(t+1) - F(t))| / \sqrt{fr}, \tag{6.1}$$

where fr is the sampling frequency, i.e., the frame rate.

ν is a ratio of the sampling rate, and it is shown to be quantitatively comparable across datasets. Specifically, a value of $\nu = 1$ indicates a very low noise level, whereas $\nu = 8$ indicates a high noise level, independently on the frame rate. Further details can be found in the supplementary material of [140].

The authors provide 27 datasets that include recordings from various brain regions of mice and zebrafish using both types of indicators. Specifically, the mouse data are from the primary visual cortex V1, the primary somatosensory cortex S1, and the CA3 region of the hippocampus, and zebrafish recordings are from the anterior and posterior parts of the telencephalic area aDp and pDp, the olfactory bulb OB, and the dorsal pallium dD. A total of 308 neural recordings, with 48 obtained from zebrafish and 260 from mice, were analyzed. The recordings vary in length, ranging from 1000 to 35,000 frames, and were sampled at different frequencies. For each dataset, if multiple recordings were available for a given neuron, the first recording is retained for further analyses. This approach ensured consistency and enabled us to perform dynamic neural noise estimation on a diverse set of neuronal activities, represented by a total of 308 series from 308 neurons.

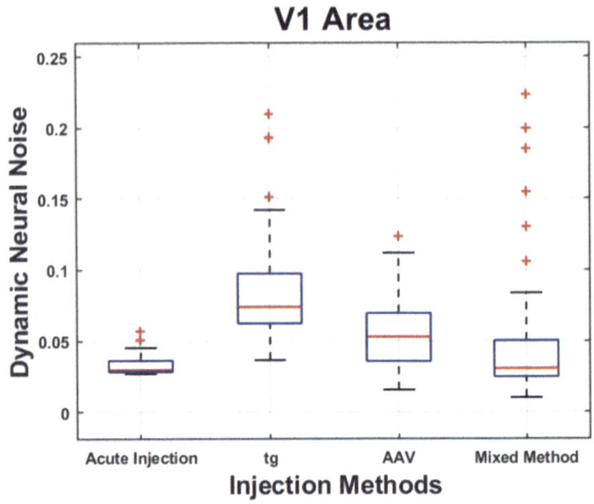

Fig. 6.14 Boxplots illustrating the distribution of dynamic neural noise levels on the V1 area of the mouse *Dataset 1* across the four injection methods (Acute, tg, AAV, and Mixed). Red lines indicate median values, while crosses indicate outliers

6.4.1.1 Results Dataset 1: Neural Spiking Activity from Calcium Imaging

First, we analyzed different labeling methods (acute injection of chemical indicators, genetically encoded indicators either through transgenic animals expressing GCaMP (tg) or virus induced expression (AAV), or labeling of transgenic animals using acute injection or AAV, referred to here as "Mixed") in the mouse cortical region V1 to investigate their contribution, if any, to region-specific noise. Region V1 was selected because it contains the largest number of neuronal recordings in Rupprecht et al.'s dataset.

The neural dynamic noise estimation always refers to the square root of the noise power within the series. As shown in the boxplot in Fig. 6.14, the tg method has the highest dynamic noise and acute injection the lowest.

The Kruskal–Wallis test confirms statistically significant differences in noise median values among the injection methods ($p \approx 10^{-5}$). Moreover, the median noise values of the labeling groups are pairwise different, while the Acute injection and Mixed dataset are not significantly different ($p = 0.77$ from pairwise multiple comparisons through Mann–Whitney tests).

Our dynamical neural noise estimates align with those from established methods for calcium imaging data. Figure 6.15 shows the correlation between estimated dynamic neural noise and fluorescence noise considering only the injecting methods, independently of the brain areas.

In zebrafish, Spearman correlation coefficients were as high as 0.923 ($p = 0.0000042$) for the acute injection case and 0.9683 ($p = 0.00000018$) for the tg(NeuroD) injection. In mouse, the Spearman correlation coefficient is 0.9541 ($p \approx 10^{-6}$) for the acute injection case and 0.9706 ($p \approx 7 \times 10^{-14}$) for the tg injection. Additionally, correlation coefficients of 0.8827 and 0.8988 ($p \approx 0$) are observed for the AAV and mixed injection procedures, respectively. The distinct correlations between the fluorescence-based noise estimations and the noise estimated using our

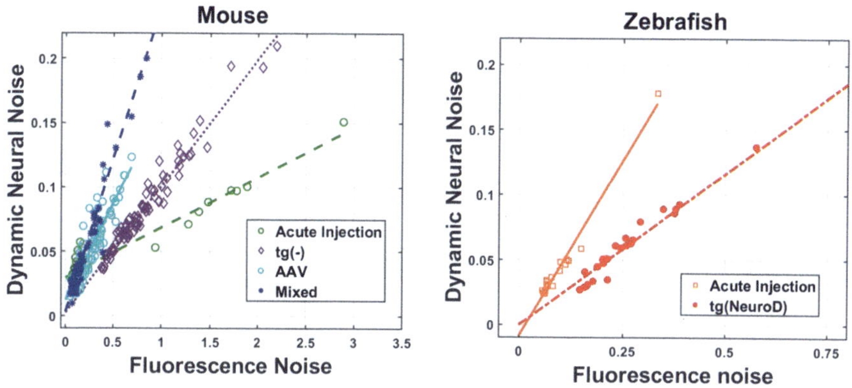

Fig. 6.15 Scatterplot illustrating the correlation between the proposed noise estimation and the fluorescence-based noise estimation for the 27 datasets from [140]. The embedding dimension is set to $m = 2$. The panel on the left specifically focuses on mouse neurons, while the right panel depicts zebrafish neurons. Neurons are color-coded and clustered based on the different injection procedures (Acute, tg, AAV, and Mixed). The lines displayed in the plot represent the linear regressions computed using the least squares method for the distinct neuronal clusters

method for different injection methods suggest that labeling methods may influence the intrinsic noise characteristics of calcium dynamics.

Lastly, Fig. 6.16 illustrates a robust and statistically significant positive correlation between the fluorescence-based formula and the proposed dynamic neural noise estimation in both zebrafish and mouse series, showing that both methods are consistent. Specifically, Spearman's correlation coefficient (ρ) between the estimations of dynamic neural noise and the fluorescence noise was greater than 0.8 and the p-value below 10^{-5}. Importantly, different linear correlation lines are observed for the three brain regions in the mouse. Similarly, the pDp region in the zebrafish brain is associated with a different linear correlation line than the other brain areas. This analysis suggests that the dynamic neural noise might contain meaningful brain area-specific information.

Furthermore, Fig. 6.17 demonstrates that the pDp area exhibits lower noise levels compared to the aDp and OB brain regions in zebrafish, as evidenced by a Mann–Whitney test (p-value < 0.01). Additionally, the dynamic neural noise level in the dorsal pallium (dD) area is significantly lower compared to the anterior part of the telencephalon (aDp) area.

6.4.2 Recordings of Cortical Activity with High Spatial Resolution from Anesthetized Rats Dataset

Neural activity is also measured using electrophysiological recordings via the patch clamp technique (usually in vitro) or using microelectrodes (in vitro or in vivo).

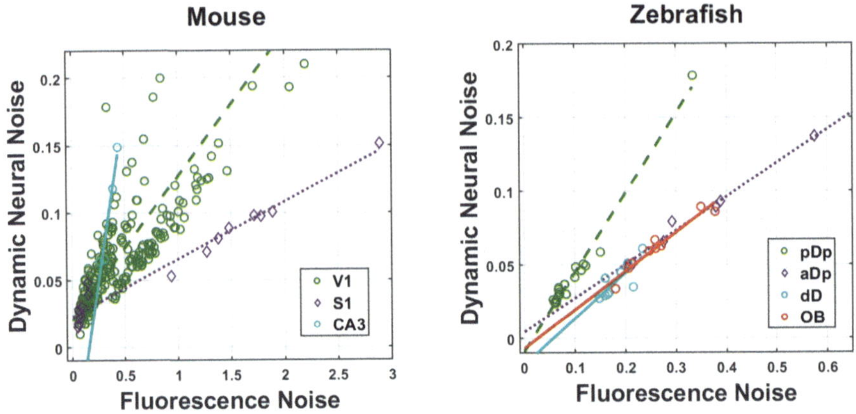

Fig. 6.16 Scatterplot depicting the correlation between the proposed noise estimation and the fluorescence-based noise estimation for the 27 datasets from [140]. The embedding dimension is set to $m = 2$. The panel on the left displays mice neurons, while the panel on the right shows zebrafish neurons. Neurons are clustered and color-coded based on their respective brain areas: primary visual cortex (V1), somatosensory cortex (S1), and region of hippocampus CA3 for mice, and anterior/posterior parts of the telencephalic area (aDp/pDp), the olfactory bulb OB, and the dorsal pallium dD for zebrafish. The lines in the plot represent the linear regressions computed via least squares for the distinct neuronal clusters

Fig. 6.17 Boxplots illustrating the distribution of dynamic neural noise levels across different brain regions for the zebrafish neurons. Neuronal recordings are clustered based on their respective brain areas, namely pDp, aDp, dD, and OB (respectively, anterior/posterior parts of the telencephalic area, the dorsal pallium dD, and the olfactory bulb)

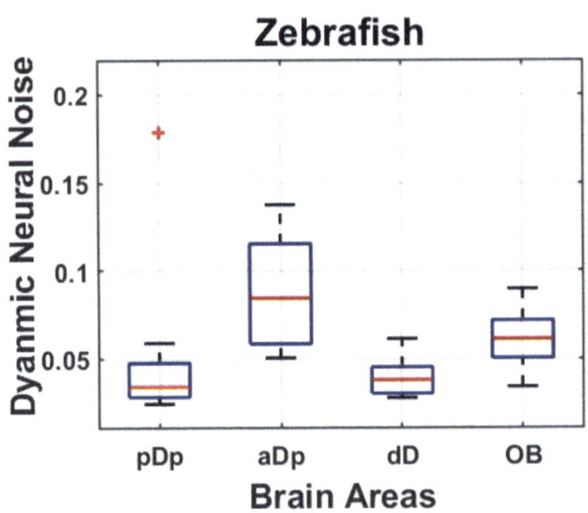

The proposed framework is used to investigate the presence of dynamic neural noise in electrophysiological recordings of cortical activity acquired from a multisite silicon-based probe implanted in anesthetized rats [66]. Data were gathered from an extensively documented dataset,[1] extensively described in [66]. In brief, cortical electrical activity was recorded using a single-shank, multisite silicon-based probe

[1] https://gin.g-node.org/UlbertLab/High_Resolution_Cortical_Spikes

equipped with 128 closely spaced titanium nitride electrodes. These recordings were conducted on 20 anesthetized rats. The recorded signals were acquired at a sampling rate of 20 kHz per channel, with a resolution of 16 bits. Approximately 30 minutes of multichannel neuronal data were collected at each recording site. We processed 39 multivariate recordings from 128 channels on a probe inserted into the superficial layers of the cortex at two different depths: "Depth1" and "Depth2." "Depth1" involved probe insertions that covered cortical activity from layers L1 (Molecular Layer), L2 (External Granular), and L3 (External Pyramidal), spanning from 0 to approximately 374 micrometers [13]. These layers consist of a small number of neurons in L1, small granule cells and interneurons in L2, and medium-sized pyramidal neurons in L3, which are involved in complex cognitive functions and associative learning [13]. Conversely, "Depth2" refers to probe insertions comprising layers L2/3, L4 (Internal Granular), and L5 (Internal Pyramidal). These layers, ranging approximately from 161 to 521 micrometers [13], are associated with sensory information processing, motor control, and action planning. Notably, layer L5 contains the largest pyramidal neurons, which project to various subcortical structures including the basal ganglia, brainstem, and spinal cord. Throughout the study, only one silicon probe was employed. Thirty-nine multivariate available recordings are processed from probe insertions into the superficial layers of the cortex. The specific areas of interest where the probes were inserted include the primary motor cortex (M1), secondary motor cortex (M2), parietal association cortex (PtA), hind limb region of the primary somatosensory cortex (S1HL), trunk region of the primary somatosensory cortex (S1Tr), anterior–posterior (AP), and medial-lateral (ML). In order to estimate dynamic neural noise, 10-second segment is extracted from each channel recording within the interval of 20 s–30 s and downsampled them at 1 kHz, resulting in 128 series of 10,000 points each for each channel recording. Thus, the overall number of processed series amounts to $39 \times 128 = 4992$ for each of the two levels. Parameter set was with $\Delta r = 0.001$ and $m = 2$.

6.4.2.1 Results Dataset 2: Cortical Activity from Anesthetized Rats

Figure 6.18 shows estimated dynamic neural noise levels across the cortical areas in anesthetized rats where the silicon probe was inserted for the level "Depth 1" and "Depth 2."

For both probe depth, significant differences were observed in the distribution of median dynamic neural noise values, measured in μV, across various brain areas, as indicated by the Kruskal–Wallis test ($p < 10^{-7}$). Further evaluation through multiple nonparametric Mann–Whitney tests confirms that the hind limb region of the primary somatosensory cortex (S1HL) consistently shows the highest level of dynamic neural noise. In contrast, the primary and secondary motor cortices, M1 and M2, respectively, demonstrate the lowest dynamic neural noise levels ($p < 10^{-7}$). Note that the levels of dynamic neural noise increase from Depth 1 to Depth 2, i.e., they increase with the depth of probe insertion, in each brain region.

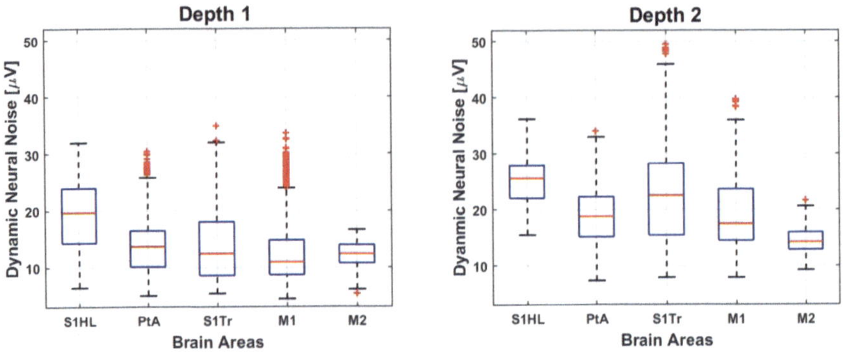

(a) Dynamic Neural Noise for Depth 1 level. (b) Dynamic Neural Noise for Depth 2 level.

Fig. 6.18 Boxplots illustrating the distribution of dynamic neural noise levels in the anesthetized rats dataset across five cortical areas (S1HL, PtA, S1Tr, M1, and M2). Panel **(a)** shows "Depth 1" probe insertion, which covers the cortical layers L1 (Molecular Layer), L2 (External Granular), and L3 (External Pyramidal), ranging from 0 to approximately 374 micrometers. Panel **(b)** displays "Depth 2" probe insertion, spanning the layers L2/3, L4 (Internal Granular), and L5 (Internal Pyramidal), with a range of approximately 161–521 micrometers. Red lines indicate median values, while crosses mark outliers

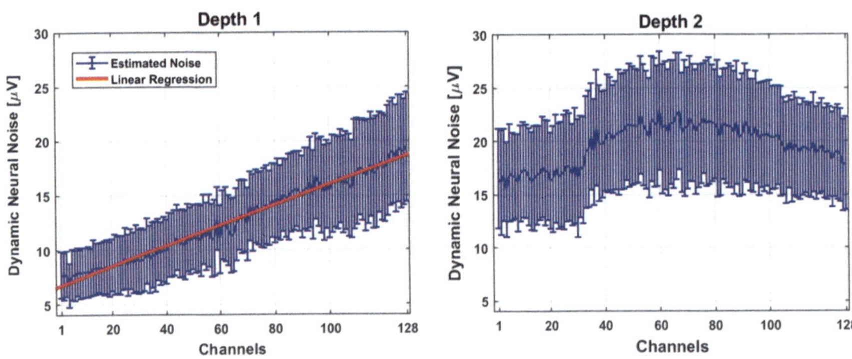

(a) (b)

Fig. 6.19 Dynamic neural noise levels (blue line), represented in terms of the median ± median absolute deviation across all 128 channels in anesthetized rats (Dataset 2), are shown for two depths of probe insertion. The left panel, "Depth 1," covers the cortical layers L1 (Molecular Layer), L2 (External Granular), and L3 (External Pyramidal), with a range extending from 0 to approximately 374 micrometers. The right panel, "Depth 2," spans the layers L2/3, L4 (Internal Granular), and L5 (Internal Pyramidal), with a range from approximately 161 to 521 micrometers. The red line represents the linear regression of the median values, with a coefficient of determination $R2 = 0.98$ for the "Depth 1" level. **(a)** Dynamic neural noise as function of probe channels for Depth 1. **(b)** Dynamic neural noise as function of probe channels for Depth 2

Figure 6.19 illustrates the channel-wise noise distribution among all rats, showing that Dynamic Neural Noise increases with probe depth until it reaches a plateau at Depth 2 (see Fig. 6.19b). Specifically, an increase in Dynamic Neural Noise is observed when transitioning from cortical layer L1 to L3, supporting the hypothesis that noise levels rise with increasing depth. Although this trend is evident across probe channels, the growth in noise ceases to be linear at cortical layers L4 and L5. This deviation may be attributed to the larger neuron sizes in these layers, which could enhance the signal-to-noise ratio and increase the deterministic component of the neural signals.

Chapter 7
Final Considerations

Abstract This chapter summarizes the key findings and outcomes derived from the physiological noise analyses performed on both synthetic and real datasets. The discussion is structured around the specific methods and their respective applications, highlighting the effectiveness and relevance of each approach.

Keywords Cardiovascular dynamics · Brain dynamics · Nonstationarity · Denoising

7.1 Physiological Noise Definition and Estimation

This book emphasizes the definition, estimation, and characterization of physiological noise in neural and cardiovascular systems.

Physiological noise is defined as the dynamical noise that impacts a system's evolution in time. Specifically, this noise is represented as a sequence of random variables, $\{\varepsilon_n\}_n$, which dynamically affects a physiological system, modeled as a dynamical system T. The influence follows the relationship $x_n = T(x_{n-1}; x_0) + \varepsilon_n$, where each x_n represents a sample point in the system starting from an initial state x_0. This framework also includes more complex relationships of the form $x_n = T(x_{n-1}, x_{n-2}, \ldots, x_0) + \varepsilon_n$.

The methodology to detect and estimate the physiological noise is based on the Approximate Entropy quantifier, which exhibits distinctive behavior under stochastic perturbations, as demonstrated in Theorem 4.1.1. Such perturbations can significantly affect complex biological systems and may bias complexity estimates (e.g., entropy); therefore the resulting dynamical noise quantification can carry valuable information on the system under study.

Note the assumption that the physiological noise $\{\varepsilon_n\}$ follows a Gaussian distribution with a mean of zero, and hence its standard deviation σ is then estimated. This modeling choice is justified by the fact that physiological noise is expected to affect all the frequencies of a physiological signal spectrum equally, rather than being concentrated in a specific frequency band (as in the case of colored noise). Although other distributions are certainly possible, the recurrent nature of

A. Scarciglia et al., *Physiological Noise and Informative Randomness in Neural and Cardiovascular Systems*, https://doi.org/10.1007/978-3-031-95914-1_7

physiological noise recursively adds up within the system. Therefore, assuming the neural noise consists of independent and identically distributed random variables—meaning the statistical distribution does not change at each iteration—the resulting sum distribution becomes "Gaussianized", in accordance with the central limit theorem.

Consequently, the proposed framework is suitable for all the physiological signals like the HRV or EEG series as it does not require any specific assumptions about the underlying dynamics, which may be constituted by unknown nonlinear functions showing complex dynamics [3, 5, 9, 47, 51, 145, 167].

However, it is important to note that the proposed framework does not depend on the specific nature or statistical distribution of the noise. In other words, the definition of noise does not necessitate prior knowledge or information, and it remains applicable regardless of the underlying dynamical complexity of the system.

Indeed, the proposed methodology provides an effective, non-heuristic estimation of dynamical noise in every kind of monodimensional time series with an unknown underlying dynamical system, like any physiological system.

In terms of noise detectability, the analysis demonstrated that noise presence in a time series results in a distinctive peak within the profile of Approximate Entropy as it varies with the tolerance parameter. Therefore, whenever a time series displays a peak in Approximate Entropy (as shown in Fig. 7.1), it can be inferred that the series is influenced by dynamical or physiological noise. The proposed methodology can then be applied to accurately estimate the power of such perturbations. This observation holds true for all the series analyzed in this book, including both synthetic and physiological signals.

On the other hand, the relevance of the noise level necessitates a more nuanced discussion. When considering noise as a measure of the complexity of a series, it is not appropriate to classify a certain level of noise as negligible. This is because the impact of intrinsic/dynamical perturbations is assessed, rather than external perturbations on the output. In situations where the encountered noise levels are significantly smaller than the range of the series, noise may produce trajectories that closely resemble the unperturbed ones, leading to the perception of negligible noise. However, in the context of intrinsic perturbations, even small disturbances can have a profound and irreversible effect on the dynamics of the series. A notable example is the impact of innovation processes on AR(p) models characterized by regular and deterministic dynamics: Even small noise variances, relative to the series range, can completely alter the resulting trajectory. Therefore, the estimated noise levels, even if it is of small magnitude, should always be considered as an indication of stochastic and irreversible changes within the underlying (unknown) dynamics.

Furthermore, when testing the proposed methodology on physiological signals (e.g., on cardiovascular and cortical activity dynamics) to investigate if physiological noise varies between systems, the results demonstrate that physiological systems are significantly driven by noise. As shown in Fig. 7.1, noise seems to change the underlying dynamics by generating peaks in the ApEn profile. Qualitatively, by comparing the four panels, it can be observed that the dynamics of HRV series

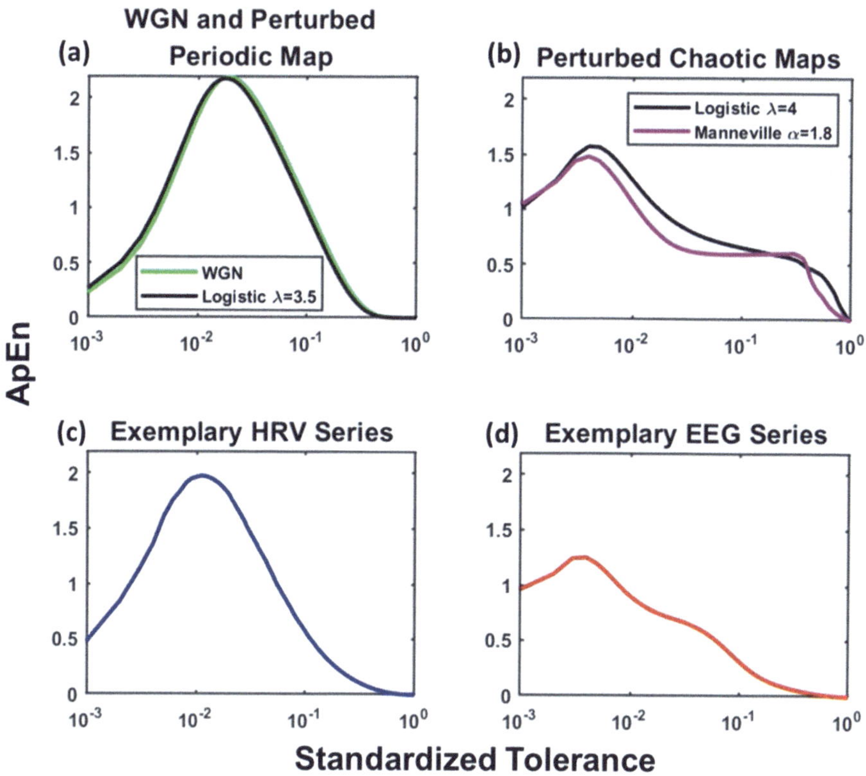

Fig. 7.1 ApEn profiles for synthetic and physiological series. The ApEn curve is represented as function of the tolerance parameter, standardized with respect to the series amplitude, with embedding dimension $m = 2$. Panel **(a)**: ApEn profiles of a realization of white Gaussian noise (WGN) and of a Logistic map in periodic regime ($\lambda = 3.5$). Panel **(b)**: ApEn profiles of a chaotic Logistic ($\lambda = 4$) and a Pomeau–Manneville map ($\alpha = 1.8$). All synthetic time series are corrupted by dynamical WGN $\sim \mathcal{N}(0, \sigma)$, being $\sigma = 5\%$ of noise-free amplitude. Panel **(c)**: ApEn profile of a time series gathered from the HRV-NS cohort. Panel **(d)**: ApEn profile of an EEG series gathered from resting state

might be more similar to a periodic series corrupted by noise (panels a–c), while the EEG ApEn profile simulates a structure more similar to chaotic systems due to the presence of a positive *plateau*, broken by the presence of noise (panels b–d).

Subsequently, an algorithm that practically estimates the physiological noise power within a signal has been described (see Chap. 4).

To evaluate the correctness and the robustness of the algorithm, the Logistic and Pomeau–Manneville maps are perturbed—in periodic and chaotic regimes with finite and infinite probability measures in the folding of the modulo 1—by imposing different levels of dynamical white Gaussian noise. Different levels of white Gaussian noise have been injected into autoregressive model of order 7, being of particular interest in computational physiology since it allows us to easily compute regularity quantifiers, such as the conditional entropy, while not restricting the dynamics range under noisy perturbation.

Numerical results reported in Chap. 5 confirm that the method properly estimates the different levels of superimposed noise in the synthetic benchmark maps. It is recalled that the standard deviation of injected noise can be up to the 20% of the signal content for the Logistic and Pomeau–Manneville maps and up to the 50% for the AR(7) model. In particular, higher variability in noise detection is associated with higher levels of injected noise.

Although the ApEn quantifier is sample-dependent [138] and the underlying theorem holds strictly for infinite-length series at any embedding dimension, the proposed method appears to be quite resilient to variations in series length. While longer time series typically provide more accurate noise estimation, reliable noise quantification is already achieved with 250–500 samples, as evidenced by the low MSE obtained. The only exception is observed in the AR(7) model, where this trend reverses—likely due to the presence of a nonstationary regime not addressed by the current framework.

Moreover, it is suggested to use a small embedding dimension m: Numerical simulations show that a higher MSE may be associated with a higher m, which would require a larger number of sample points to properly estimate conditional probabilities in higher dimensional space.

Regarding the method sensitivity, the method significantly recognizes linear variations of dynamical noise; for what concerns the choice of the resolution step Δr, it is recommended to use $1/1000 \times$ {time series range} as a good trade-off between accuracy and computational cost.

The method applies to any continuous and differentiable T map, and no further *a priori* information on T is required. Furthermore, the proposed method is effective for the estimation of additive (output) noise, too. While the method does not provide information on the nature of the noise (output or dynamical) or its statistical properties, such estimation may have practical advantages in real applications. In fact, no a priori information or hypotheses are necessary to perform the proposed noise power estimation.

The fact that the proposed method is applicable to both discrete-time and continuous-time dynamical systems deserves attention. The noise effects are considered on dynamics between t_1 and t_2 as the amount of noise carried by a continuous-time dynamical system at the instant t_2. This sum of noise effects is then replaced with a discrete-time noise deviation, which can be numerically handled in observable, sampled (and hence discrete-time) time series. Therefore, it can be assumed that the observable series are the output of a discrete-time dynamical system. This is particularly relevant for applications involving, e.g., financial, natural, and complex physiological systems, since their recordings are discrete-time signals by definition, being the result of sampling.

Indeed, in the neurocardiovascular system, a link between the level of noise in neural circuits and the activity of the norepinephrine system, which is part of the sympathetic nervous system, has been established [130]. It is not surprising that such sympathetic activity and, thus, neurocardiac noise may be modulated by cognitive states [130], as the sympathetic system is part of the autonomic nervous system that innervates all body organs and systems, including the cardiovascular system. This neural control of cardiac activity is part of the central autonomic network, which means that noise may affect the dynamics of the brain and body as a whole, including interconnected networks [188, 190]. The combined neural control of heartbeat activity through acetylcholine, cytosolic adenosine 3,5-cyclic monophosphate, and other mediators of postjunctional interactions between the sympathetic and vagal systems may result in neurocardiovascular noise [167]. For what concerns cortical dynamics, it is known that EEG originates from the activity of pyramidal neurons in the cortex and seems to be associated with a relatively high physiological noise level, whose distribution varies across the scalp. Multiple factors may influence its presence, including but not limited to, neurovascular coupling, neuro-glia interactions, and related calcium dynamics, as well as vascular motion due to heartbeat dynamics.

7.2 Cardiovascular Dynamics

Physiological Noise and Aging Previous studies have suggested that complexity in heartbeat dynamics decreases with age [159, 177]; however, such findings may be biased by the implicit assumption of a noise-free cardiovascular system. Further analyses presented in Chap. 6 reveal that aging is associated with reduced levels of physiological noise in autonomic control of heartbeat dynamics, as evidenced by a lower noise standard deviation and reduced intersubject variability in elderly individuals.

These observations support the hypothesis that aging might primarily influence physiological noise, rather than directly altering the underlying system dynamics. This interpretation aligns with prior evidence suggesting that aging does not significantly affect the complexity of cardiovascular dynamics [152]. Consequently, it may be inferred that the presence of physiological noise—largely unaccounted for in earlier complexity analyses—may have substantially confounded prior conclusions regarding age-related changes in cardiovascular complexity [3, 4, 9, 51, 135, 145].

Physiological Noise and Pathology The results presented in Chap. 6 indicate that cardiovascular noise is significantly influenced by cardiac pathological conditions compared to the healthy state. At the group level, physiological noise accounts for approximately 5–10% of heartbeat dynamics—corresponding to about 30–40% of the overall signal standard deviation—and is notably modulated by disease.

Specifically, conditions such as atrial fibrillation and congestive heart failure are associated with markedly noisier cardiac dynamics than those observed in healthy controls, who also exhibit considerably lower intersubject variability. These findings suggest that cardiac pathology may primarily modulate the level of intrinsic physiological noise, rather than altering the complexity of the underlying system dynamics.

This interpretation aligns with prior studies reporting reduced signal-to-noise ratios and dynamics resembling white Gaussian noise in cardiac disorders [32]. These observations highlight the need for further investigation into the reliability of complexity-based metrics—such as entropy—as indicators of cardiovascular regulation across different experimental and clinical contexts [185].

7.3 Cortical Dynamics

With regard to cortical dynamics, the physiological noise investigation reveals a distinct spatial distribution across the scalp (see Fig. 6.7). At the group level, this noise accounts for approximately 1.5–2% of the EEG dynamics, corresponding to about 8–11% of the total signal standard deviation, and appears to be modulated by mental states. Similar to the spatial distribution of EEG alpha power in the resting state, higher physiological noise levels are observed in occipital regions, while central electrodes exhibit lower levels. Notably, significant differences in noise distribution emerge across experimental conditions.

During the mental arithmetic task, elevated noise levels are detected in posterior scalp areas and left prefrontal regions compared to the resting state. This suggests that cognitive engagement during the task modulates not only cortical activity but also the underlying neurophysiological noise, with specific effects in prefrontal and occipital brain regions.

The observed increase in dynamical noise during the mental arithmetic task may be physiologically linked to fluctuations in cerebral hemoglobin oxygenation during cognitive processing [129] or to the suppression of alpha rhythms in the parieto-occipital regions [107]. Speculatively, the marked reduction in alpha activity may unmask an underlying aperiodic and noisy component that is otherwise obscured by dominant rhythmic patterns.

In EEG analysis, the ability to distinguish signal from noise is critical for accurate interpretation of brain activity. While several physiological and nonphysiological noise sources have been identified, the distinction between signal and noise is often subjective and context-dependent, influenced by current theoretical and methodological perspectives on brain function [103]. Indeed, throughout the history of neurophysiological research, components initially classified as noise have, in some cases, been reinterpreted as meaningful signals of interest [103].

7.4 Brain Dynamics

Physiological noise in brain dynamics might stem from the stochastic nature of various biological processes, such as gene expression, cellular activity, or neural fluctuations. Assuming that measurement noise is uncorrelated with aging due to its distribution across data acquisition from various subjects and MRI machines, intrinsic physiological noise in the brain is modulated with age progression in healthy individuals. This conclusion is based on results from the Cam-CAN and HCP Connectome 1200 datasets, which reveal a distinct spatial distribution of physiological noise components in brain activity during the resting state, along with age-related modulation of their power. Across both datasets, approximately 80–85% of the regions of interest (ROIs) exhibit similar noise behavior. Specifically, most ROIs show a statistically significant increase in noise levels with aging. In the Cam-CAN dataset, only two ROIs exhibit a negative trend, while no significant negative trends are observed in the HCP dataset, likely due to the narrower age range of the participants.

Validation of the method on synthetic fMRI signals with known noise characteristics supports the robustness of the observed positive trend. Although differences in series length could introduce bias, the increase in noise was independently confirmed across datasets. The most strongly correlated regions' activity with physiological noise included the right frontal operculum, the orbital part of the right inferior frontal gyrus, precentral areas, Rolandic operculum, medial cingulum, and occipital regions. Age-related modulation of fMRI noise was consistently observed through both standard ROI-based analyses and ICA spatial clustering in two independent, publicly available, preprocessed datasets. Although ICA spatial clustering is often employed as a denoising technique, the dynamical noise identified in this study appears to be intrinsic to the brain's functional activity. This makes it fundamentally different from removable artifacts and suggests that its elimination is not straightforward, as it may carry physiologically meaningful information.

7.5 Neuronal Dynamics

Intrinsic, dynamic neural noise (i.e., physiological noise in neural systems), which can be estimated directly from neural time series, can be investigated using calcium imaging or invasive neural recordings. Before estimating dynamic neural noise, the method's reliability was assessed using synthetic data generated from both Izhikevich's neuronal model and the calcium diffusion model for different noise levels (see Chap. 5). Results demonstrated that the methods described in Chap. 4 effectively identify the different levels of injected dynamical noise in such synthetic data.

When analyzing experimental neural datasets, variations in dynamical neural noise levels occur across different brain regions and among the different calcium indicator labeling methods.

Of note, when comparing calcium imaging data from different mouse brain regions, no differences of dynamic neural noise were observed among the $V1$, $S1$, and $CA3$ regions. This result might be a consequence of the small number of neural recordings of the $S1$ and $CA3$ areas with respect to the $V1$ area. But, when considering only the V1 area for the mouse recordings, the labeling seems to elicit higher levels of dynamic noise with respect to the acute or mixed method injections.

On the other hand, in the zebrafish the pDp and dD areas showed lower dynamic neural noise, as well as lower fluorescence noise levels, compared to adD and OB.

Moreover, a strong correlation is found between the dynamic neural noise estimation and the fluorescence noise method applied to the calcium imaging dataset.

While fluorescence noise estimation is inherently constrained by the physical characteristics of the calcium imaging process, the dynamic neural noise estimation proposed herein is formulated within a more generalized theoretical framework. This broader approach extends its applicability to a wider range of physiological data related to neural activity, including those analyzed in this book.

Neuronal recordings from anesthetized rats showed that dynamic neural noise is predominant in the hind limb region of the primary somatosensory cortex. In contrast, the primary (M1) and secondary (M2) motor cortices were associated with lower dynamic neural noise power.

The primary motor (M1/M2) and somatosensory (S1) cortices are frequently studied because they are well-mapped regions with strong input/output connectivity, making physiological effects—such as changes in background synaptic activity, local connectivity, or modulatory input—more identifiable and quantifiable. Our findings demonstrate that the somatosensory hind limb region (S1HL) exhibits higher neural noise, likely due to increased synaptic input and integration, local complexity, and state-dependent activity. Specifically, S1HL receives continuous ascending sensory input, leading to elevated spontaneous synaptic activity compared to motor cortices, which, at rest, experience fewer concurrent inputs. Thalamocortical loops projecting to the somatosensory cortex contribute to greater signal variability and stochastic fluctuations [60]. Moreover, S1HL contains diverse populations of inhibitory interneurons and excitatory projection neurons, forming a dense network in which recurrent connections amplify random fluctuations, increasing overall dynamic neural noise. Finally, even under anesthesia, somatosensory areas receiving strong bottom-up input can exhibit burst-suppression patterns or spontaneous oscillations, superimposing an additional "noisy" baseline compared to motor regions [60, 97].

The laminar organization of the cortex likely also plays a role in dynamic neural noise levels. In particular, deeper cortical layers contain extensive recurrent collateral networks and subcortical projections, which contribute to network-driven fluctuations [97, 143]. Additionally, neuromodulatory inputs from thalamic or basal forebrain sources vary by depth, and the summation of excitatory and inhibitory postsynaptic potentials in deeper layers introduces high-frequency stochastic components into electrophysiological signals [60]. Additional explanations for the increase in Dynamic Neural Noise from cortical layers L1–L3, and its subsequent cessation at layers L4 and L5, may be linked to augmented thermal effects on neurons as a function of depth during brain stimulation [87]. These thermal effects can complicate the detection of spike and neuronal firing events [141, 197], making it challenging to accurately identify neural activities at greater depths.

Results on the different labeling methods raise interesting questions because the standard deviation of dynamic neural noise appears to depend on the method of labeling. Notably, transgenic mice expressing GCaMP6 have higher levels of dynamic noise than mice injected with Calcium dyes such as OGB-1 and Ca-520, despite the fact that genetically encoded calcium indicators are generally preferred for their better and more consistent signal-to-noise ratio over time, along with their ability to be genetically targeted, which adds to their specificity and reduces background noise. The significance and impact of this bias should be investigated in further studies [175].

Regarding the observation that GECIs exhibit higher dynamic neural noise compared to acutely injected calcium dyes, the underlying cause likely may arise from a combination of biological and optical factors. The viral (AAV) or transgenic expression of GECIs depends on promoter strength, cell type, and protein expression variability. Uneven distribution within the cytosol can introduce fluctuations in baseline fluorescence and calcium buffering [175]. Some GECIs have slower on/off kinetics compared to chemical dyes, causing calcium transients to accumulate fluctuations over time, which appear as increased noise [27]. While the observation of higher noise in GECIs is consistent with findings from multiple laboratories—where GECIs generally provide a lower initial SNR but greater flexibility for chronic imaging—a rigorous distinction between biological variability (e.g., differences in calcium handling, expression levels) and technical artifacts (e.g., baseline brightness, kinetics, and imaging parameters) is crucial. Future studies should compare well-matched GECI- and dye-labeled animals while controlling for imaging depth, indicator concentration, excitation intensity, and animal health status. Ultimately, the evidence of increased dynamic noise in certain cortical regions or when using specific indicators underscores the importance of combining multi-region, layer-resolved recordings with rigorous physiological and technical controls.

Overall, these findings contribute to the field of computational biology by introducing a reliable method for estimating dynamic neural noise and highlighting the impact of the injection procedure on noise levels in neuronal data. The identified noise is an intrinsic neural characteristic that complements other important transient calcium features, such as the duration and amplitude of the signal [69] and related detection methods [83]

These findings emphasize the importance of carefully considering noise sources and their potential effects in neuroscientific studies. Future research should be addressed toward additional factors contributing to noise in neural data and investigate potential methods for noise reduction or mitigation. Furthermore, the relationship between noise levels and specific experimental parameters, such as electrode placement or recording duration, could be further probed to gain a comprehensive understanding of the factors influencing noise in neural recordings.

7.6 Nonstationarity of Physiological Noise

A method for detecting physiological noise nonstationarity in complex physiological time series is presented in Chap. 4. The proposed framework treats physiological noise as a recursive realization of IID random variables, typically Gaussian $\mathcal{N}(0, \sigma)$. This is achieved through a window-sliding process, which returns a sample-wise physiological noise power estimation $\psi_{RR}(t', \mathcal{H}_{t'}, \varepsilon(t'))$ and its variability throughout the entire series. The method has been tested on four different classes of synthetic time series and applied it to real cardiovascular data gathered from a publicly available postural changes dataset to detect possible variations during significant sympathovagal changes, recognizing the fundamental, intrinsic role of stochastic dynamics in the cardiovascular system.

For synthetic data (see Chap. 5), the method effectively differentiated different levels of superimposed dynamical noise using a relatively short window length of only 100 samples. This is notable because the noise estimation algorithm [149] theoretically requires a large number of samples to perform accurately. The accuracy of sample-wise noise estimation along the series improves with larger time windows, as evidenced by decreased dispersion around the median value, resulting from a more precise characterization of the approximate entropy quantifier. Moreover, for periodic dynamics and autoregressive models, the sample-wise noise estimation appears extremely accurate. In contrast, a slight overestimation of noise levels is present for chaotic data, probably due to the need for more samples to properly distinguish the dynamics from the noise presence. In smaller sample observations, the series might appear more random, obscuring the underlying dynamical pattern.

A separate discussion is necessary for IPFM HRV model, as its formulation inherently involves stochastic perturbations interacting with a dynamical component or multiple integration processes. Indeed, the noise observed in IPFM HRV models tends to be significantly underestimated. This discrepancy may be attributed to the unique nature of the IPFM model, which generates heartbeat series by integrating an input signal until it reaches a predetermined threshold parameter. When noise is introduced into the integration process, predicting how the noise samples interact becomes challenging. There is a possibility that they cancel each other out, given their mean is 0, as per the assumptions. Nonetheless, the proposed method distinctly discern between different levels of dynamical noise in a time-resolved way.

All classical time and frequency domain quantifiers exhibit a similar trend to our time-varying noise estimate, which is expected given the relationship between noise and signal variability. However, our method, which is less sensitive to underlying dynamics, reveals that changes in time and frequency domain parameters, commonly associated with sympathovagal and vagal activity, may be attributed to stochastic fluctuations rather than neural mechanisms. This has also been further demonstrated by the synthetic data analysis shown in Fig. 5.13.

The proposed method finds a natural application on the HRV time series gathered from the postural changes dataset, since both the PP and KPSS tests confirmed the nonstationarity of the physiological time series undergoing the different tasks. Results showed that physiological noise in the cardiovascular system tends to increase from supine rest to upright standing. This increase is more pronounced during the first 60s of stand-up or fast tilt transitions from rest. Subsequent changes show a significant decrease in physiological noise levels with respect to the supine resting state. Normalized physiological noise dynamics significantly decreases after posture changes. Specifically, the proportion of heartbeat dynamics explained by noise is around 60–70% during rest and decreases to 30–40% in the upright position. Interestingly, the transition from rest to slow tilt is associated with a minimum physiological noise approximately 120 seconds after the tilt onset, while the minimum is reached around 60 seconds after the fast tilt and stand-up tasks.

These findings demonstrate that physiological noise is not merely a constant in the dynamics, but rather an intrinsic and nonstationary component of the system—as confirmed by statistical tests and by the time-dependent frequency modulation in the spectrograms of the estimated noise time series—thus variable and informative of the system's state.

At a speculative level, the results suggest that the intrinsic stochasticity in cardiovascular dynamics is primarily driven by an increase in sympathetic activity coupled with a withdrawal of vagal activity, as absolute physiological noise trends mirror those of sympathetic activity [186]. Conversely, normalized noise dynamics appear to follow vagal activity [186]. Notably, the reduction in vagal and normalized noise due to postural changes is consistent with earlier research showing reduced cardiovascular complexity during posture changes [187].

From a biochemical perspective, while its precise origin remains elusive, this nonstationary physiological noise likely arises from the complex interplay within the autonomic nervous system [85]—cardiovascular dynamics exhibit inherent nonlinearity and are influenced by physiological randomness. Neurotransmitters like norepinephrine, released by the sympathetic nervous system, and acetylcholine, released by the parasympathetic system, along with other mediators, contribute to this stochasticity [6, 130, 167].

The findings suggest that previous entropy-based complexity assessments of HRV series may be influenced by natural complexity changes associated with noise [148].

The limitations of this analysis include a reduced number of subjects and trials, as well as the theoretical challenge of determining the optimal sliding window size for more precise noise estimation, which would mitigate the impact of transition phases between two consecutive noise powers of nonstationary noise.

7.7 Physiological Denoising

A novel model-free method is described in Chap. 4 aiming to reduce the dynamical noise levels in a regular, nonlinear time series. It relies on state-space reconstruction and time-reversing, one-step forecasting, selecting the best value in a neighborhood of the multidimensional space. The method's philosophy is to smooth the current value of a noisy time series using the value that generates the highest number of orbits close to each other, despite the noise drift. It is supposed that dynamical noise is a sequence of IID Gaussian variables with mean 0 and variance σ^2 that dynamically corrupts the time series evolution.

For the selection of neighborhood amplitude in the multidimensional space where the method operates, it is suggested to use twice the intrinsic physiological noise standard deviation detected through the framework for monovariate noise estimation, since the maximal noise reduction is achieved in a neighborhood of such value for almost any map. This choice of the neighborhood size in the embedded state space seems more convenient than the time-consuming refining of coarse-grained time space partition proposed in [155]. Indeed, the proposed method only applies one time the denoising procedure, while the literature technique needs to repeat the noise reduction steps for each value of the neighborhood amplitude refined grid. A formal justification of the neighborhood value 2σ as a reliable threshold can be addressed to the possibility of finding almost all the noise drifts—for a sequence of IID random variables with distribution $\mathcal{N}(0, \sigma)$, 95% of the values are in the range $(-2\sigma, 2\sigma)$.

Numerical simulations on synthetic time series in Chap. 5 with different levels of injected noise have confirmed the satisfactory performance of the proposed denoising with respect to the reference Schreiber's method: The proposed method performs better for nonlinear, regular maps when the injected noise is not significantly high. Longer series surely enhance the method's ability to find similar patterns and hence achieve better noise removal.

When dealing with pure Gaussian noise realizations, no peculiar improvements are reported from our method with respect to the reference one, probably due to the absence of dynamics which does not allow retrieving recurrent noisy patterns for denoising. The method seems to perform better with higher threshold values than the suggested one, as larger neighborhoods allow for a more accurate estimation of the current value through the average of the samples. This is because the statistical mean serves as an unbiased estimator for the mean value of IID white Gaussian noise processes. Furthermore, it is speculated that the optimal result is not achieved with a tolerance value close to the suggested one due to the complete randomness and lack of dynamical structure in the noisy process or because $m = 3$ is not the right embedding dimension for a stochastic process—if it makes sense to define a state space for a random process. Nonetheless, in case of pure white noise series, the proposed noise reduction method removes up to 75% of the injected noise, and its performance does not differ significantly from the existing Schreiber's method. This minor difference can be attributed to the fact that the final samples of the noisy time series are not smoothed when using the proposed method.

Regarding logistic maps, it has been observed that the proposed noise reduction method outperformed the reference method, achieving a lower RMSE and reaching optimal denoising near the suggested threshold. Even modest improvements are significant in removing dynamical noise, as small perturbations can significantly alter the system's dynamics. With an increased number of samples, lower noise levels enhance denoising effectiveness, halving the dynamical error for lower values and eliminating up to 25% of the noise for higher imposed values. In cases of higher noise levels, a greater dynamical error is noted, likely due to the modulo 1 reduction used in defining the map, which introduces more significant perturbations in the series.

For Rössler maps, the suggested value for achieving the maximum noise reduction seems to increase along with the number of samples. This can be justified by the fact that noise perturbs the whole three-dimensional trajectory, even if it has been superimposed on only one single axis: Hence, the total amount of noise on one axis is actually increased along with the multiple integration steps, due to the nonlinearity of Eq. 5.9.

For the suggested tolerance values, a significant reduction has been observed. Despite the fact that the SNR of the introduced perturbations is not high, this small stochasticity still influences the integration of the dynamics.

The problem of dynamical noise reduction is still an issue when only a small number of samples can be collected from experimental measurement, and it deserves further investigation. Furthermore, for the neighborhood amplitude suggested by [149], Schreiber's method [155] also seems to assume the minimum for noise reduction. Hence, the latter can be more efficient and less time-consuming by using the model-free noise estimation.

The noise reduction on real HRV datasets demonstrates the significant impact of dynamical physiological noise on complexity analysis, potentially biasing signal interpretation. Our proposed method is particularly suitable for heartbeat dynamics series, as the cardiovascular system can be reasonably modeled by a periodic noisy map. Recent findings indicate that these HRV series are not generated by chaotic maps but by regular, potentially nonlinear dynamics with high levels of dynamical noise [15]. After physiological denoising, the median noise and entropy levels of healthy subjects and those with atrial fibrillation show minimal differences. This aligns with prior research [148], suggesting that noise is an intrinsic feature of the system. The modulation of physiological noise, considered as an external perturbation to cardiac dynamics, correlates with different physiological conditions. Its removal results in more homogeneous dynamics across various physiological states. In essence, assuming the underlying dynamics remain consistent across different cohorts, stochasticity differentiates cardiac states, with its absence leading to similar entropy and reduced noise levels. This finding is consistent with studies indicating that the complexity, and dynamics of cardiac cycles are stable across various diseases and aging [152, 183], while differing complexity behaviors arise from varying levels of stochastic perturbations. Additionally, the results corroborate previous research [32] showing that healthy, CHF, and AF cohorts exhibit different levels of complexity, as measured by entropy across multiple scales. Specifically, multiscale sample entropy analysis revealed that at the smallest time scale, sample entropy for the healthy cohort was lower compared to the other two cohorts, with this trend reversing at higher time scales, indicating that healthy individuals exhibit the highest cardiovascular complexity. This change in complexity is further supported by other metrics such as symbolic entropy [7]. Along with these quantifiers, the proposed method reduces the LF and HF spectral powers, as expected by subtracting part of the noise, which is white. Moreover, the noise reduction method seems not to perturb statistical quantifiers such as the mean—and slightly affect variance—of the signal probably because each point of the cleaned series is obtained by searching for an average of the values which are really close in a small neighborhood of the actual, noise-free series.

7.8 Final Remarks and Conclusion

Noise has played a foundational role in the development of applied sciences. Whether considered as interference or as an intrinsic component of a system, its presence can significantly bias signal analysis by altering both the statistical and dynamical properties of the system—often increasing its complexity and introducing apparent chaotic behavior.

Traditionally, noise has been regarded as an artifact affecting signal acquisition and experimental reproducibility. However, a more contemporary perspective recognizes (dynamical) noise as an integral feature of biological and physiological systems, with a fundamental role in modeling and understanding various physiological phenomena. When noise is treated as an external disturbance—commonly referred to as measurement or output noise—numerous techniques exist for its identification and removal. However, intrinsic or dynamical noise, which is embedded within the system and influences its behavior at each time step, presents a far more complex challenge. This is particularly true when the governing equations of the system are unknown. Methods for identifying such intrinsic physiological noise primarily rely on three theoretical frameworks: (1) dynamical systems theory, using complexity quantifiers such as entropy measures, (2) probability theory and Bayesian inference, and (3) stochastic differential equations, often used in state-space models like the Kalman filter, which assume at least partial knowledge of the system's structure. This challenge becomes especially relevant in systems where the underlying dynamics are poorly understood, such as heart rate variability (HRV), electroencephalography (EEG), calcium imaging, neuronal spike recordings, and functional MRI (fMRI). These signals reflect processes at various physiological scales—from cellular ion exchange and neural spiking to autonomic and cortical activity. This book aims to define and quantitatively estimate physiological noise in complex biological systems without relying on explicit analytical models of their dynamics. To this end, we adopted a hybrid framework that captures both deterministic and stochastic features of physiological systems. Each system is modeled as a deterministic process perturbed by an intrinsic stochastic component—dynamical noise—which is not simply an artifact, but a consistent influence that alters the system's trajectory. The methodology described in Chap. 4 centers on the divergence behavior of a nonlinear complexity measure—Approximate Entropy (ApEn)—in the presence of stochastic components, regardless of the system's underlying dynamics. This property forms the basis of algorithms developed to estimate noise levels in physiological time series. The proposed approach was validated on synthetic datasets (see Chap. 5) generated from well-known dynamical systems and information-theoretic models and then extended to real-world physiological signals (see Chap. 6). These analyses revealed that physiological noise is measurable and exhibits distinct physiological and pathophysiological signatures, influencing both cardiovascular and neural systems. For instance, noise in EEG signals is spatially heterogeneous and can be modulated by cognitive tasks, such as mental arithmetic. In such cases, increased noise was observed in specific brain regions, suggesting that task-driven modulation targets both system dynamics and the embedded stochastic components.

Accurate estimation of physiological noise is thus essential for evaluating physiological function and pathophysiological states. Many commonly used physiological metrics, including power spectral density and entropy, are highly sensitive to noise. Analyses of physiological data that omit quantitative noise assessment can lead to distorted or misleading results.

Physiological noise in both cortical and cardiovascular systems is modulated by aging. Specifically, normalized noise accounts for up to 70% of variability in HRV and approximately 23% in EEG signals. Aging is associated with decreased neurocardiovascular noise and increased neurocortical noise, particularly in frontal and occipital areas. These results suggest that aging influences physiological systems not only through anatomical adaptations but also by modulating their stochastic components. This challenges prior interpretations of complexity metrics in aging and emphasizes the need to reconsider existing results in light of physiological noise.

Regarding brain activity, a consistent increase in stochastic noise with age was observed across most regions. This modulation can lead to misinterpretations of biomarkers if noise is not accurately accounted for—potentially affecting diagnostic or clinical decisions when comparing biomarker values across brain areas.

In this context, the presented methodology offers a novel tool for estimating dynamical noise from real-world neural signals. Results from both synthetic and experimental datasets demonstrate the method's ability to quantify noise levels and differentiate between neuronal behaviors and brain regions. This opens new opportunities for refining models and analytical strategies in neuroscience.

The analysis also addresses the nonstationarity of physiological noise. In cardiovascular systems, fluctuations in noise levels were significantly associated with changes in physiological conditions.

Additionally, we introduced a model-free method for reducing dynamical noise in regular, nonlinear time series (see Chap. 4). This approach employs state-space reconstruction, time-reversed one-step forecasting, and an orbit-based estimation of the most likely true value, assuming independent and identically distributed (i.i.d.) Gaussian noise. A key innovation of the method is the optimal setting of the neighborhood amplitude to twice the estimated noise standard deviation, enhancing denoising performance across a variety of dynamical maps. Our simulations show this method outperforms existing approaches, particularly for regular systems with moderate noise, and proves robust when applied to HRV datasets—enabling more accurate complexity assessments.

Finally, we demonstrate that physiological noise can distort the coupling dynamics in networks, such as in the Kuramoto oscillator model. This finding has important implications for studying interconnected physiological systems, such as cortical and brain networks, where neglecting noise may result in incorrect inferences about connectivity and coordination.

Despite the demonstrated effectiveness and broad applicability of the proposed methodologies for estimating physiological noise, the estimation algorithm requires sufficiently long time series. When data are too short, they offer only a localized view of the underlying physiological phenomena, limiting the method's ability to capture both the deterministic dynamics and the stochastic components effectively. This constraint is especially relevant in time-varying noise estimation, where rapidly changing noise levels are difficult to resolve over short observation windows.

Regarding the denoising approach, two key limitations emerge: the length of the time series and the selection of the embedding dimension in the Takens' reconstructed space. For the method to be effective, the time series must be long enough to contain repeated or comparable patterns, which are necessary for the identification of close orbits within the reconstructed state space. Additionally, accurately determining the optimal embedding dimension is crucial for capturing the system's relevant dynamics. While embedding theorems offer theoretical lower bounds to ensure topological equivalence of the reconstructed dynamics, such guarantees do not extend to short or noisy time series. Thus, the choice of embedding parameters remains heuristic and must be adapted to the available data. To improve robustness, this method may be complemented by alternative denoising techniques, such as those based on spectral analysis or model-based approaches.

Future research directions should be directed toward the refinement of noise estimation methods for very short time series—an important step toward enabling real-time or bedside applications. From an applied perspective, physiological noise should be tested on further datasets to evaluate its role as a potential biomarker in a broader range of pathophysiological conditions, including stress, neurodegenerative diseases (e.g., Alzheimer's), and in response to specific sensory–motor or cognitive stimuli (e.g., visual, auditory, and motor tasks). Further investigations will also explore stochastic influences in intersystem interactions, such as neural—motor or cardio—respiratory coupling. Another important direction involves advancing the biological and physiological understanding of the mechanisms modulated by intrinsic noise and identifying the underlying triggers of such modulation.

From a methodological standpoint, integrating the physiological noise estimation framework with information-theoretic approaches will be a valuable area of research. While information theory offers powerful tools for quantifying the informational content of physiological signals—typically treated as realizations of random variables—an exclusively stochastic view may obscure the contribution of deterministic dynamics versus stochastic fluctuations. As noise is often associated with increased information or complexity, the proposed methodology could provide a means of disentangling the informational contribution of intrinsic noise from the total information content of the system.

Finally, a promising area of future study refers to causality estimation techniques, such as Granger causality, Convergent Cross Mapping, and Transfer Entropy. Since noise can distort both geometric and statistical properties of time series, it may obscure or misrepresent underlying causal relationships. The noise estimation techniques developed in this book could offer valuable tools to quantify the extent to which stochastic components influence or mask causality, helping to correct or reinterpret causal inferences in complex physiological systems.

Chapter 8
Appendix

Proof of Theorem 4.1.1 Without loss of generality, we suppose $X = [0, 1]$, and we consider $|\cdot|$ as the infinity norm.

Case $m = 1$. Observe that for any admissible integer k,

$$|x_{n+k+1} - x_{n+1}| = |T(x_{n+k}) - T(x_n) + \varepsilon_{n+k+1} - \varepsilon_{n+1}|,$$

where ε_{n+1} and ε_{n+k+1} are independent noise samples.

Let $\psi(t)$ denote the PDF of the random variable $\zeta = \varepsilon_{n+k+1} - \varepsilon_{n+1}$, and let $\mathrm{d}\mu(x) = h(x)\mathrm{d}x$ be a density function that is C^1.

We estimate the conditional probability $\mathcal{P}$ of having any two series iterates x_{n+1} and x_{n+k+1} r-close, given x_n and x_{n+k} r-close for any admissible integer k. By setting $z := x_n$ and $y := x_{n+h}$, the conditional probability P can be expressed as follows: $\int \frac{\int_{z-r}^{z+r} \int_{T(z)-T(y)-r}^{T(z)-T(y)+r} \psi(t)\mathrm{d}t \, \mathrm{d}\mu(y)}{\int_{z-r}^{z+r} 1 \, \mathrm{d}\mu(y)} \mathrm{d}\mu(z)$. Since the approximate entropy is the negative expectation of the log probability $\mathcal{P}$, we obtain

$$\begin{aligned} &\mathrm{ApEn}\left(\{x_n(\xi)\}_{n=1}^{\infty}, 1, r\right) = \\ &- \int \log \frac{\int_{z-r}^{z+r} \int_{T(z)-T(y)-r}^{T(z)-T(y)+r} \psi(t)\mathrm{d}t \, \mathrm{d}\mu(y)}{\int_{z-r}^{z+r} 1 \, \mathrm{d}\mu(y)} \mathrm{d}\mu(z). \end{aligned} \tag{8.1}$$

Next, we formalize the changes of ApEn when the radius r becomes comparable to noise perturbation effects. Take a fixed value of z, and consider J_z to be a small neighborhood centered at 0. Let $N_z(r) = \int_{z-r}^{z+r} \int_{\alpha(y,r)}^{\beta(y,r)} \psi(t)\mathrm{d}t \, \mathrm{d}\mu(y)$ for $r \in J_z$, with $\alpha(y, r) := T(z) - T(y) - r$ and $\beta(y, r) := T(z) - T(y) + r$. We have $N_z(0) = 0$.

A. Scarciglia et al., *Physiological Noise and Informative Randomness in Neural and Cardiovascular Systems*, https://doi.org/10.1007/978-3-031-95914-1_8

As the limit bounds of integration and integrand functions in $N_z(r)$ are differentiable by thesis, we can apply Leibniz's rule to obtain the following result:

$$N_z'(r) = \int_{z-r}^{z+r} [\psi(\beta(y,r)) + \psi(\alpha(y,r))]h(y)\mathrm{d}y +$$

$$\int_{\alpha(z+r,r)}^{\beta(z+r,r)} h(z+r)\psi(t)\mathrm{d}t + \int_{\alpha(z-r,r)}^{\beta(z-r,r)} h(z-r)\psi(t)\mathrm{d}t.$$

$N_z'(r)$ is continuous in J_z, and we still have $N'(0) = 0$.

If $\psi(t)$ is differentiable, then integrand functions and bounds of integration in $N'(r)$ are differentiable: The Leibniz rule yields

$$\begin{aligned} N_z''(r) = {} & h(z+r)(1 - T'(z+r))(\psi(\beta(z+r,r)))+ \\ & h(z+r)(1 + T'(z+r))(\psi(\alpha(z+r,r)))+ \end{aligned}$$

$$\int_{\alpha(z+r,r)}^{\beta(z+r,r)} h'(z+r)\psi(t)\mathrm{d}t - \int_{\alpha(z-r,r)}^{\beta(z-r,r)} h'(z-r)\psi(t)\mathrm{d}t +$$

$$\begin{aligned} & h(z-r)(1 + T'(z-r))(\psi(\beta(z-r,r))) + \\ & -h(z-r)(-1 + T'(z-r))(\psi(\alpha(z-r,r)))+ \\ & h(z+r)(\psi(\beta(z+r,r)) + \psi(\alpha(z+r,r)))+ \\ & h(z-r)(\psi(\beta(z-r,r)) + \psi(\alpha(z-r,r)))+ \\ & \int_{z-r}^{z+r} h(y)[\psi'(\beta(y,r)) - \psi'(\alpha(y,r))]\mathrm{d}y. \end{aligned}$$

Since $N_z''(0) = 8h(z)\psi(0)$ is defined and $N'(r)$ is C^1, we can use the Taylor polynomial expansion $N_z(r) = 4h(z)\psi(0)r^2 + o(r^2)$ in J_z.

If $\psi(t)$ is not differentiable at $t = 0$, we can still obtain $N_z''(0) = 8h(z)\psi(0)$ by using the smoothing of ψ through mollifiers, which is a sequence of C^∞ functions that uniformly converges to ψ in any small neighborhood of 0.

Moreover, $\int_{z-r}^{z+r} 1\mathrm{d}\mu(y) \approx 2rh(z)$ in J_z.

By substituting the previous polynomial expansions into (8.2), we can obtain the desired result since, for $r > 0$,

$$\begin{aligned} \text{ApEn}\left(\{x_n(\xi)\}_{n=1}^{\infty}, 1, r\right) &\approx -\int \log \frac{4h(z)\psi(0)r^2}{2rh(z)} \mathrm{d}\mu(z) \\ &= -\int \log\left[2\psi(0)r\right]\mathrm{d}\mu(z) = -\log 2\psi(0)r. \end{aligned} \tag{8.2}$$

Case $m > 1$. We embed the series $\{x_n(\xi)\}_{n=1}^{\infty}$ in an m-dimensional space by forming vectors $\mathbf{X}_j = (x_{j-(m-1)}, \ldots, x_j)$ for all admissible j. Analogously, $\text{ApEn}(\{x_n(\xi)\}_{n=1}^{\infty}, m, r)$ is minus the expectation of the log probability $\mathcal{P}$ of having two iterates x_{n+k+1} and x_{n+1} r-close, being the previous m points $\mathbf{X}_{n+k}$ $\mathbf{X}_n$ r-close for any admissible integer k.

The proof directly retraces the case $m = 1$: The only exception is of considering the invariant measure μ defined on an m-dimensional space. □

Proof of Corollary 4.1.2 If the noise samples are IID Gaussian distributed random variables of mean 0 and variance σ^2, then the density function $\psi \sim \mathcal{N}(0, 2\sigma^2)$.

In this context,

$$\psi(0) = \frac{1}{2\sigma\sqrt{\pi}},$$

and hence relation (4.1) becomes

$$\text{ApEn}\left(\{x_n(\sigma)\}_{n=1}^{\infty}, m, r\right) \approx -\log\left[r/(\sigma\sqrt{\pi})\right].$$

□

Bibliography

1. https://archive.physionet.org/physiobank/database/html/mitdbdir/mitdbdir.htm
2. Aaronson, J. (1997). *An introduction to infinite ergodic theory*. Number 50. American Mathematical Society.
3. Acharya, U. R., et al. (2006). Heart rate variability: A review. *Medical and Biological Engineering and Computing, 44*(12), 1031–1051.
4. Alù, F., Orticoni, A., Judica, E., Cotelli, M., Rossini, P. M., Miraglia, F., & Vecchio, F. (2021). Entropy modulation of electroencephalographic signals in physiological aging. *Mechanisms of Ageing and Development, 196*, 111472.
5. Armoundas, A. A., Ju, K., Iyengar, N., Kanters, J. K., Saul, P. J., Cohen, R. J., & Chon, K. H. (2002). A stochastic nonlinear autoregressive algorithm reflects nonlinear dynamics of heart-rate fluctuations. *Annals of Biomedical Engineering, 30*(2), 192–201.
6. Aston-Jones, G., & Cohen, J. D. (2005). An integrative theory of locus coeruleus-norepinephrine function: adaptive gain and optimal performance. *Annual Review of Neuroscience, 28*(1), 403–450.
7. Aziz, W., Rafique, M., Ahmad, I., Arif, M., Habib, N., & Nadeem, M. (2014). Classification of heart rate signals of healthy and pathological subjects using threshold based symbolic entropy. *Acta Biologica Hungarica, 65*(3):252–264.
8. Bachelier, L. (1900). Théorie de la spéculation. In *Annales scientifiques de l'École normale supérieure* (Vol. 17, pp. 21–86).
9. Barbieri, R., et al. (2017). *Complexity and nonlinearity in cardiovascular signals*. Springer.
10. Barrett, A. B., & Seth, A. K. (2011). Practical measures of integrated information for time-series data. *PLoS Computational Biology, 7*(1), e1001052.
11. Bergemann, K., & Reich, S. (2012). An ensemble kalman-bucy filter for continuous data assimilation. *Meteorologische Zeitschrift, 21*(3):213.
12. Bernido, C. C., & Carpio-Bernido, M. V. (2014). *Methods and applications of white noise analysis in interdisciplinary sciences*. World Scientific.
13. Bertero, A., Verrillo, L., & Apicella, A. j. (2022). A novel layer 4 corticofugal cell type/projection involved in thalamo-cortico-striatal sensory processing. *Journal of Neuroscience, 42*(8), 1383–1405.
14. Bianchi, F., Formentin, S., & Piroddi, L. (2020). Process noise covariance estimation via stochastic approximation. *International Journal of Adaptive Control and Signal Processing, 34*(1):63–76. Cited by: 5; All Open Access, Green Open Access.
15. Bianco, M., Scarciglia, A., Bonanno, C., & Valenza, G. (2025). Quantifying chaotic behavior in noisy dynamical systems: A study on heartbeat dynamics. *IEEE Transactions on Biomedical Engineering*, 1–10. https://doi.org/10.1109/TBME.2025.3566470

A. Scarciglia et al., *Physiological Noise and Informative Randomness in Neural and Cardiovascular Systems*, https://doi.org/10.1007/978-3-031-95914-1

16. Bilgin, N., & Olivier, A. (2025). Joint bayesian estimation of process and measurement noise statistics in nonlinear kalman filtering. *Mechanical Systems and Signal Processing, 223*. Cited by: 0.
17. Blinks, J. R., Prendergast, F. G., & Allen, D. G. (1976). Photoproteins as biological calcium indicators. *Miscellaneous, 28*(1), 93.
18. Bonanno, C., & Lenci, M. (2021). Pomeau-manneville maps are global-local mixing. *Discrete & Continuous Dynamical Systems, 41*(3), 1051.
19. Boschi, C. D. E., Ortega, G. J., & Louis, E. (2002). Discriminating dynamical from additive noise in the Van der Pol oscillator. *Physica D: Nonlinear Phenomena, 171*(1–2), 8–18.
20. Böttcher, F., Peinke, J., Kleinhans, D., Friedrich, R., Lind, P. G., & Haase, M. (2006). Reconstruction of complex dynamical systems affected by strong measurement noise. *Physical Review Letters, 97*(9), 090603.
21. Brennan, M., Palaniswami, M., & Kamen, P. (2002). Poincare plot interpretation using a physiological model of HRV based on a network of oscillators. *American Journal of Physiology-Heart and Circulatory Physiology, 283*(5), H1873–H1886.
22. Brin, M., & Stuck, G. (2002). *Introduction to dynamical systems*. Cambridge University Press.
23. Brodal, P. (2010). *The central nervous system*. Oxford University Press.
24. Bryant, H. L., & Segundo, J. P. (1976). Spike initiation by transmembrane current: A white-noise analysis. *The Journal of physiology, 260*(2), 279–314.
25. Cabral, J., Hugues, E., Sporns, O., & Deco, G. (2011). Role of local network oscillations in resting-state functional connectivity. *Neuroimage, 57*(1), 130–139.
26. Carnevale, N. T., & Hines, M. L. (2006). *The NEURON book*. Cambridge University Press.
27. Chen, T.-W., Wardill, T. J., Sun, Y., Pulver, S. R., Renninger, S. L., Baohan, A., Schreiter, E. R., Kerr, R. A., Orger, M. B., Jayaraman, V., et al. (2013). Ultrasensitive fluorescent proteins for imaging neuronal activity. *Nature, 499*(7458), 295–300.
28. Citi, L., Brown, E. N., & Barbieri, R. (2012). A real-time automated point-process method for the detection and correction of erroneous and ectopic heartbeats. *IEEE Transactions on Biomedical Engineering, 59*(10), 2828–2837.
29. Çoban, G., Büyüklü, A. H., & Das, A. (2012). A linearization based non-iterative approach to measure the Gaussian noise level for chaotic time series. *Chaos, Solitons & Fractals, 45*(3), 266–278.
30. Cohen, L. (2005). The history of noise [on the 100th anniversary of its birth]. *IEEE Signal Processing Magazine, 22*(6), 20–45.
31. Cohen, M. X. (2017). Where does EEG come from and what does it mean? *Trends in Neurosciences, 40*(4), 208–218.
32. Costa, M., Goldberger, A. L., & Peng, C.-K. (2005). Multiscale entropy analysis of biological signals. *Physical review E, 71*(2), 021906.
33. Da Silva, F. L. (2023). EEG: Origin and measurement. In *EEG-fMRI: physiological basis, technique, and applications* (pp. 23–48). Springer.
34. de C Hamilton, A. F., Jones, K. E., & Wolpert, D. M. (2004). The scaling of motor noise with muscle strength and motor unit number in humans. *Experimental Brain Research, 157*, 417–430.
35. Denk, G., & Winkler, R. (2007). Modelling and simulation of transient noise in circuit simulation. *Mathematical and Computer Modelling of Dynamical Systems, 13*(4):383–394.
36. Eckmann, J.-P., & Ruelle, D. (1985). Ergodic theory of chaos and strange attractors. *Reviews of Modern Physics, 57*(3):617.
37. Einstein, A. (1905). Über die von der molekularkinetischen theorie der wärme geforderte bewegung von in ruhenden flüssigkeiten suspendierten teilchen. *Annalen der physik, 322*(8):549–560. https://doi.org/10.1002/andp.19053220806
38. Faisal, A. A., Selen, L. P. J., & Wolpert, D. M. (2008). Noise in the nervous system. *Nature Reviews Neuroscience, 9*(4):292–303.
39. Falk, J., Mendler, M., & Drossel, B. (2017). A minimal model of burst-noise induced bistability. *PloS One, 12*(4), e0176410.

40. Fatt, P., & Katz, B. (1950). Some observations on biological noise. *Nature, 166*(4223), 597–598.
41. Faure, P., & Korn, H. (1998). A new method to estimate the kolmogorov entropy from recurrence plots: Its application to neuronal signals. *Physica D: Nonlinear Phenomena, 122*(1–4), 265–279.
42. Fellous, J.-M., Rudolph, M., Destexhe, A., & Sejnowski, T. J. (2003). Synaptic background noise controls the input/output characteristics of single cells in an in vitro model of in vivo activity. *Neuroscience, 122*(3), 811–829.
43. Friedman, A. (2005). Introduction to neurons. In *Tutorials in Mathematical Biosciences I: Mathematical Neuroscience* (pp. 1–20). Springer. https://doi.org/10.1007/978-3-540-31544-5_1
44. Gao, X., Luo, H., Ning, B., Zhao, F., Bao, L., Gong, Y., Xiao, Y., & Jiang, J. (2020). RL-AKF: An adaptive Kalman filter navigation algorithm based on reinforcement learning for ground vehicles. *Remote Sensing, 12*(11). Cited by: 46; All Open Access, Gold Open Access.
45. Geerligs, L., Henson, R. N., et al. (2016). Functional connectivity and structural covariance between regions of interest can be measured more accurately using multivariate distance correlation. *NeuroImage, 135*, 16–31.
46. Gillespie, D. T., et al. (1996). The mathematics of Brownian motion and Johnson noise. *American Journal of Physics, 64*(3), 225–239.
47. Glass, L. (2009). Introduction to controversial topics in nonlinear science: Is the normal heart rate chaotic? *Chaos: An Interdisciplinary Journal of Nonlinear Science, 19*(2), 028501. https://doi.org/10.1063/1.3156832
48. Glover, G. H. (1999). Deconvolution of impulse response in event-related BOLD fMRI1. *Neuroimage, 9*(4), 416–429.
49. Goldberg, B.-G. (2000). Phase noise theory and measurements: A short review. *Microwave Journal, 43*(1), 112–112.
50. Goldberger, A. L., Amaral, L. A. N., Glass, L., Hausdorff, J. M., Ivanov, P. Ch., Mark, R. G., Mietus, J. E., Moody, G. B., Peng, C.-K., & Stanley, H. E. (2000). Physiobank, physiotoolkit, and physionet: Components of a new research resource for complex physiologic signals. *Circulation, 101*(23), e215–e220.
51. Goldberger, A. L., et al. (2002). What is physiologic complexity and how does it change with aging and disease? *Neurobiol Aging, 23*(1), 23–26.
52. Gorban, I. (2012). Statistically instable processes: Connection with flicker, nonequilibrium, fractal and colored noise. *Radioelectronics and Communications Systems, 55*(3), 99–114.
53. Grassberger, P., & Procaccia, I. (1983). Estimation of the kolmogorov entropy from a chaotic signal. *Physical Review A, 28*(4), 2591.
54. Grassberger, P., & Procaccia, I. (1983). Measuring the strangeness of strange attractors. *Physica D: Nonlinear Phenomena, 9*(1–2), 189–208.
55. Gray, R. M. (1990). Quantization noise spectra. *IEEE Transactions on Information Theory, 36*(6), 1220–1244.
56. Graybiel, A. M. (2000). The basal ganglia. *Current Biology, 10*(14), R509–R511.
57. Grienberger, C., & Konnerth, A. (2012). Imaging calcium in neurons. *Neuron, 73*(5), 862–885.
58. Hall, P., & Wood, A. (1993). On the performance of box-counting estimators of fractal dimension. *Biometrika, 80*(1), 246–251.
59. Hamilton, F., Berry, T., & Sauer, T. (2017). Kalman-takens filtering in the presence of dynamical noise. *The European Physical Journal Special Topics, 226*, 3239–3250.
60. Harris, K. D., & Thiele, A. (2011). Cortical state and attention. *Nature Reviews Neuroscience, 12*(9), 509–523.
61. Heald, J. P. M., & Stark, J. (2000). Estimation of noise levels for models of chaotic dynamical systems. *Physical Review Letters 84*,(11), 2366.
62. Heldt, T., Oefinger, M. B., Hoshiyama, M., & Mark, R. G. (2003). Circulatory response to passive and active changes in posture. In *Computers in cardiology, 2003* (pp. 263–266). IEEE.

63. Higuchi, T. (1988). Approach to an irregular time series on the basis of the fractal theory. *Physica D: Nonlinear Phenomena, 31*(2), 277–283.
64. Hirata, Y. (2018). Reconstructing latent dynamical noise for better forecasting observables. *Chaos: An Interdisciplinary Journal of Nonlinear Science, 28*(3), 033112. https://doi.org/10.1063/1.4996043
65. Holm, S. (1979). A simple sequentially rejective multiple test procedure. *Scandinavian Journal of Statistics, 6*(2):65–70. http://www.jstor.org/stable/4615733
66. Horváth, C., Tóth, L. F., Ulbert, I., & Fiáth, R. (2021). Dataset of cortical activity recorded with high spatial resolution from anesthetized rats. *Scientific Data, 8*(1), 180.
67. Hu, G., Gao, B., Zhong, Y., & Gu, C., (2020). Unscented Kalman filter with process noise covariance estimation for vehicular INS/GPS integration system. *Information Fusion, 64*, 194–204. Cited by: 140
68. Hubbard, J. (2012). *The peripheral nervous system*. Springer Science & Business Media.
69. Huerta, T. S., Haider, B., Adamovich-Zeitlin, R., Chen, A. C., Chaudhry, S., Zanos, T. P., Chavan, S. S., Tracey, K. J., & Chang, E. H. (2023). Calcium imaging and analysis of the jugular-nodose ganglia enables identification of distinct vagal sensory neuron subsets. *Journal of Neural Engineering, 20*(2), 026014.
70. Itoh, K.-i. (1996). A method for detecting impulsive noises in chaotic time series. *IEICE TRANSACTIONS on Fundamentals of Electronics, Communications and Computer Sciences, 79*(10), 1670–1675.
71. Iyengar, N., et al. (1996). Age-related alterations in the fractal scaling of cardiac interbeat interval dynamics. *American Journal of Physiology-Regulatory, Integrative and Comparative Physiology, 271*(4), R1078–R1084.
72. Izhikevich, E. M. (2003). Simple model of spiking neurons. *IEEE Transactions on Neural Networks, 14*(6), 1569–1572.
73. Izhikevich, E. M. (2004). Which model to use for cortical spiking neurons? *IEEE Transactions on Neural Networks, 15*(5), 1063–1070.
74. Jackson, A. F., & Bolger, D. J. (2014). The neurophysiological bases of EEG and EEG measurement: A review for the rest of us. *Psychophysiology, 51*(11), 1061–1071.
75. Jayawardena, A. W., Penchang, Xu., & Sivakumar, B. (2002). Noise level estimation of chaotic hydrological time series. *International Association of Hydrological Sciences, Publication, 274*, 297–306. https://www.scopus.com/inward/record.uri?eid=2-s2.0-0036094505&partnerID=40&md5=e8db836da4629e4bec01bf10bed6e601
76. Jayawardena, A. W., Xu, P., & Li, W. K. (2008). A method of estimating the noise level in a chaotic time series. *Chaos: An Interdisciplinary Journal of Nonlinear Science, 18*(2), 023115.
77. Jessen, K. R. (2004). Glial cells. *The International Journal of Biochemistry & Cell Biology, 36*(10), 1861–1867.
78. Jiang, H., & Cai, Y. (2018). Adaptive fifth-degree cubature information filter for multi-sensor bearings-only tracking. *Sensors, 18*(10). https://doi.org/10.3390/s18103241. https://www.mdpi.com/1424-8220/18/10/3241
79. Jones, K. E., de C Hamilton, A. F., & Wolpert, D. M. (2002). Sources of signal-dependent noise during isometric force production. *Journal of Neurophysiology, 88*(3), 1533–1544.
80. Kalman, R. E. (1960). A new approach to linear filtering and prediction problems. *Journal of Basic Engineering, 82*(1), 35–45. https://doi.org/10.1115/1.3662552
81. Kalnay, E. (2003). *Atmospheric modeling, data assimilation and predictability*. Cambridge University Press.
82. Kaloudis, K., & Hatjispyros, S. J. (2018). A bayesian nonparametric approach to dynamical noise reduction. *Chaos: An Interdisciplinary Journal of Nonlinear Science, 28*(6), 063110. https://doi.org/10.1063/1.5025545
83. Kang, S., Park, J., Kim, K., Lim, S.-H., Kim, S., Choi, J. H., Rah, J.-C., & Choi, J.-W. (2022). ICoRD: Iterative correlation-based ROI detection method for the extraction of neural signals in calcium imaging. *Journal of Neural Engineering, 19*(4), 046035.
84. Kantz, H., & Schreiber, T. (2004). *Nonlinear time series analysis* (Vol. 7). Cambridge University Press.

85. Karavaev, A. S., Ishbulatov, Y. M., Ponomarenko, V. I., Bezruchko, B. P., Kiselev, A. R., & Prokhorov, M. D. (2019). Autonomic control is a source of dynamical chaos in the cardiovascular system. *Chaos: An Interdisciplinary Journal of Nonlinear Science, 29*(12), 121101. https://doi.org/10.1063/1.5134833
86. Katok, A., Katok, A. B., & Hasselblatt, B. (1995). *Introduction to the modern theory of dynamical systems*. Number 54. Cambridge University Press.
87. Kim, T. K., Kadji, H., Whalen, A. J., Ashourvan, A., Freeman, E., Fried, S. I., Tadigadapa, S., & Schiff, S. J. (2022). Thermal effects on neurons during stimulation of the brain. *Journal of Neural Engineering, 19*(5), 056029.
88. Koopman, F. A., Stoof, S. P., Straub, R. H., Van Maanen, M. A., Vervoordeldonk, M. J., & Tak, P. P. (2011). Restoring the balance of the autonomic nervous system as an innovative approach to the treatment of Rheumatoid Arthritis. *Molecular Medicine, 17*, 937–948.
89. Krum, H., Bigger, T., Jr., Goldsmith, R. L., & Packer, M. (1995). Effect of long-term digoxin therapy on autonomic function in patients with chronic heart failure. *Journal of the American College of Cardiology, 25*(2), 289–294.
90. Kugiumtzis, D. (1997). Correction of the correlation dimension for noisy time series. *International Journal of Bifurcation and Chaos, 7*(06), 1283–1294.
91. Kuo, H.-H. (2018). *White noise distribution theory*. CRC Press.
92. Kuramoto, Y., & Kuramoto, Y. (1984). *Chemical turbulence*. Springer.
93. Kwan, A. C. (2008). What can population calcium imaging tell us about neural circuits? *Journal of Neurophysiology, 100*(6), 2977–2980.
94. Lai, Z., Lei, Y., Zhu, S., Xu, Y.-L., Zhang, X.-H., & Krishnaswamy, S. (2016). Moving-window extended Kalman filter for structural damage detection with unknown process and measurement noises. *Measurement: Journal of the International Measurement Confederation, 88*, 428–440. Cited by: 33.
95. Langevin, P., et al. (1908). Sur la théorie du mouvement brownien. *Comptes rendus de l'Académie des Sciences Paris, 146*(530–533), 530.
96. Lax, M., Cai, W., & Xu, M. (2006). *Random processes in physics and finance*. Oxford University Press.
97. Lefort, S., Tomm, C., Floyd Sarria, J.-C., & Petersen, C. C. H. (2009). The excitatory neuronal network of the C2 barrel column in mouse primary somatosensory cortex. *Neuron, 61*(2), 301–316.
98. Leontitsis, A. (2006). Red noise estimation. *International Journal of Bifurcation and Chaos, 16*(07), 2111–2117.
99. Leontitsis, A., Pange, J., & Bountis, T. (2003). Large noise level estimation. *International Journal of Bifurcation and Chaos, 13*(08), 2309–2313.
100. Levy, M. N. (1971). Brief reviews: Sympathetic-parasympathetic interactions in the heart. *Circulation Research, 29*(5), 437–445.
101. Li, J., Fan, W., Li, Y., Yang, B., & Lu, C. (2020). Desert seismic noise suppression based on an improved low-rank matrix approximation method. *Journal of Applied Geophysics, 173*, 103926. https://doi.org/10.1016/j.jappgeo.2019.103926
102. Li, Y.-X., & Rinzel, J. (1994). Equations for InsP3 receptor-mediated [Ca2+] i oscillations derived from a detailed kinetic model: A hodgkin-huxley like formalism. *Journal of theoretical Biology, 166*(4), 461–473.
103. Liu, T. T. (2016). Noise contributions to the fMRI signal: An overview. *NeuroImage, 143*, 141–151.
104. Lizier, J. T. (2012). *The local information dynamics of distributed computation in complex systems*. Springer Science & Business Media.
105. Logothetis, N. K., Pauls, J., Augath, M., Trinath, T., & Oeltermann, A. (2001). Neurophysiological investigation of the basis of the fMRI signal. *Nature, 412*(6843), 150–157.
106. MacDonald, D. K. C. (2006). *Noise and fluctuations: an introduction*. Courier Corporation.
107. Magosso, E., De Crescenzio, F., Ricci, G., Piastra, S., & Ursino, M. (2019). EEG alpha power is modulated by attentional changes during cognitive tasks and virtual reality immersion. *Computational Intelligence and Neuroscience, 2019*(1), 7051079.

108. Mai, J. K., & Paxinos, G. (2011). *The human nervous system*. Academic Press.
109. Mandelbrot, B. B. (1982). *The fractal geometry of nature*. WH Freeman and Co.
110. Mañé, R. (1981). A proof of Pesin's formula. *Ergodic Theory and Dynamical Systems, 1*(1):95–102.
111. Manneville, P., & Pomeau, Y. (1980). Different ways to turbulence in dissipative dynamical systems. *Physica D: Nonlinear Phenomena, 1*(2), 219–226.
112. Marieb, E. N., & Hoehn, K. (2007). *Human anatomy & physiology*. Pearson Education.
113. Martínez-Rey, M., Santos, C., Nieto, R., Losada, C., & Espinosa, F. (2019). Online evaluation of the process noise covariance matrix for event-based state estimators. *International Journal for Numerical Methods in Engineering, 120*(2), 195–208. Cited by: 0.
114. Mediano, P. A. M., Rosas, F., Carhart-Harris, R. L., Seth, A. K., & Barrett, A. B. (2019). Beyond integrated information: A taxonomy of information dynamics phenomena. Preprint. arXiv:1909.02297.
115. Mediano, P. A. M., Rosas, F. E., Farah, J. C., Shanahan, M., Bor, D., & Barrett, A. B. (2022). Integrated information as a common signature of dynamical and information-processing complexity. *Chaos: An Interdisciplinary Journal of Nonlinear Science, 32*(1), 013115. https://doi.org/10.1063/5.0063384
116. Michael-Titus, A. T., & Shortland, P. (2022). *The nervous system, e-book: The nervous system, e-book*. Elsevier Health Sciences.
117. Molnár, Z., Clowry, G. J., Šestan, N., Alzu'bi, A., Bakken, T., Hevner, R. F., Hüppi, P. S., Kostović, I., Rakic, P., Anton, E. S., et al. (2019). New insights into the development of the human cerebral cortex. *Journal of Anatomy, 235*(3), 432–451.
118. Moody, G. B., & Mark, R. G. (1983). A new method for detecting atrial fibrillation using R-R intervals. *Computers in Cardiology, 1983*(10), 227–230.
119. Moyal, J. E. (1949). Quantum mechanics as a statistical theory. In *Mathematical Proceedings of the Cambridge Philosophical Society* (Vol. 45, pp. 99–124). Cambridge University Press.
120. Nadkarni, S., & Jung, P. (2004). Dressed neurons: Modeling neural–glial interactions. *Physical Biology, 1*(1):35.
121. Nguyen, J. P., Shipley, F. B., Linder, A. N., Plummer, G. S., Liu, M., Setru, S. U., Shaevitz, J. W., & Leifer, A. M. (2016). Whole-brain calcium imaging with cellular resolution in freely behaving caenorhabditis elegans. *Proceedings of the National Academy of Sciences, 113*(8), E1074–E1081.
122. Noakes, L. (1991). The takens embedding theorem. *International Journal of Bifurcation and Chaos, 1*(04), 867–872.
123. Noble, A., Johnson, R., Thomas, A., & Bass, P. (2013). The cardiovascular system e-book: The cardiovascular system e-book. In *Systems of the Body*, Churchill Livingstone. https://books.google.it/books?id=T1gnAgAAQBAJ
124. Nolte, G., Ziehe, A., & Müller, K.-R. (2001). Noise robust estimates of correlation dimension and K 2 entropy. *Physical Review E, 64*(1), 016112.
125. Ohkura, M., Matsuzaki, M., Kasai, H., Imoto, K., & Nakai, J., (2005). Genetically encoded bright Ca2+ probe applicable for dynamic Ca2+ imaging of dendritic spines. *Analytical Chemistry, 77*(18), 5861–5869.
126. Packard, N. H., Crutchfield, J. P., Farmer, J. D., & Shaw, R. S. (1980). Geometry from a time series. *Physical Review Letters, 45*, 712–716.
127. Pan, J., & Tompkins, W. J. (1985). A real-time qrs detection algorithm. *IEEE Transactions on Biomedical Engineering, BME-32*(3), 230–236. https://doi.org/10.1109/TBME.1985.325532
128. Parpura, V., Heneka, M. T., Montana, V., Oliet, S. H. R., Schousboe, A., Haydon, P. G., Stout, R. F. Jr, Spray, D. C., Reichenbach, A., Pannicke, T., et al. (2012). Glial cells in (patho) physiology. *Journal of Neurochemistry, 121*(1), 4–27.
129. Pathan, N. S., Foysal, M., & Alam, Md.M. (2019). Efficient mental arithmetic task classification using wavelet domain statistical features and svm classifier. In *2019 International Conference on Electrical, Computer and Communication Engineering (ECCE)* (pp. 1–5). IEEE.

130. Pertermann, M., Mückschel, M., Adelhöfer, N., Ziemssen, T., & Beste, C. (2019). On the interrelation of 1/f neural noise and norepinephrine system activity during motor response inhibition. *Journal of Neurophysiology, 121*(5), 1633–1643.
131. Pikovsky, A., Rosenblum, M., & Kurths, J. (2001). Synchronization. *Cambridge University Press* (Vol. 12). Cambridge university press.
132. Pincus, S. M. (1991). Approximate entropy as a measure of system complexity. *Proceedings of the National Academy of Sciences, 88*(6), 2297–2301.
133. Porta, A., De Maria, B., Bari, V., Marchi, A., & Faes, L. (2016). Are nonlinear model-free conditional entropy approaches for the assessment of cardiac control complexity superior to the linear model-based one? *IEEE Transactions on Biomedical Engineering, 64*(6), 1287–1296.
134. Porta, A., Gnecchi-Ruscone, T., Tobaldini, E., Guzzetti, S., Furlan, R., & Montano, N. (2007). Progressive decrease of heart period variability entropy-based complexity during graded head-up tilt. *Journal of Applied Physiology, 103*(4), 1143–1149.
135. Rhea, C. K., et al. (2011). Noise and complexity in human postural control: interpreting the different estimations of entropy. *PloS One, 6*(3), e17696.
136. Rice, S. O. (1944). Mathematical analysis of random noise. *The Bell System Technical Journal, 23*(3), 282–332.
137. Richman, J. S., Lake, D. E., & Moorman, J. R. (2004). Sample entropy. In *Methods in enzymology* (Vol. 384, pp. 172–184). Elsevier.
138. Richman, J. S., & Moorman, J. R. (2000). Physiological time-series analysis using approximate entropy and sample entropy. *American Journal of Physiology-Heart and Circulatory Physiology, 278*(6), H2039–H2049. PMID: 10843903.
139. Robbins, S. M., Christensen, C., Kaminski, C., & Zlatic, M. (2021). Calcium imaging analysis–how far have we come? *F1000Research, 10*, 258. https://doi.org/10.12688/f1000research.51755.2
140. Rupprecht, P., Carta, S., Hoffmann, A., Echizen, M., Blot, A., Kwan, A. C., Dan, Y., Hofer, S. B., Kitamura, K., Helmchen, F., et al. (2021). A database and deep learning toolbox for noise-optimized, generalized spike inference from calcium imaging. *Nature Neuroscience, 24*(9), 1324–1337.
141. Sadras, N., Pesaran, B., & Shanechi, M. M. (2024). Shanechi, event detection and classification from multimodal time series with application to neural data. *Journal of Neural Engineering, 21*(2), 026049.
142. Salinet, J. L., Jr, & Silva, O. L. (2019). ECG signal acquisition systems. In *Developments and applications for ECG signal processing* (pp. 29–51). Elsevier.
143. Santuy, A., Tomás-Roca, L., Rodríguez, J.-R., González-Soriano, J., Zhu, F., Qiu, Z., Grant, S. G. N., DeFelipe, J., & Merchan-Perez, A. (2020). Estimation of the number of synapses in the hippocampus and brain-wide by volume electron microscopy and genetic labeling. *Scientific Reports, 10*(1), 14014.
144. Sase, T., Ramírez, J. P., Kitajo, K., Aihara, K., & Hirata, Y. (2016). Estimating the level of dynamical noise in time series by using fractal dimensions. *Physics Letters A, 380*(11–12), 1151–1163.
145. Sassi, R., et al. (2015). Advances in heart rate variability signal analysis: joint position statement by the e-cardiology esc working group and the european heart rhythm association co-endorsed by the asia pacific heart rhythm society. *EP Europace, 17*(9), 1341–1353.
146. Sauer, T., Yorke, J. A., & Casdagli, M. (1991). Embedology. *Journal of Statistical Physics, 65*, 579–616.
147. Scarciglia, A., Catrambone, V., Bianco, M., Bonanno, C., Toschi, N., & Valenza, G. (2024). Stochastic brain dynamics exhibits differential regional distribution and maturation-related changes. *NeuroImage, 290*, 120562.
148. Scarciglia, A., Catrambone, V., Bonanno, C., & Valenza, G. (2024). Physiological noise: Definition, estimation, and characterization in complex biomedical signals. *IEEE Transactions on Biomedical Engineering, 71*(1), 45–55. https://doi.org/10.1109/TBME.2023.3291538

149. Scarciglia, A., Gini, F., Catrambone, V., Bonanno, C., & Valenza, G. (2023). Estimation of dynamical noise power in unknown systems. *IEEE Signal Processing Letters, 30*, 234–238.
150. Schartner, M., Seth, A., Noirhomme, Q., Boly, M., Bruno, M.-A., Laureys, S., & Barrett, A. (2015). Complexity of multi-dimensional spontaneous EEG decreases during propofol induced general anaesthesia. *PloS One, 10*(8), e0133532.
151. Schleicher, D. (2007). Hausdorff dimension, its properties, and its surprises. *The American Mathematical Monthly, 114*(6), 509–528.
152. Schmitt, D. T., & Ivanov, P. Ch. (2007). Fractal scale-invariant and nonlinear properties of cardiac dynamics remain stable with advanced age: a new mechanistic picture of cardiac control in healthy elderly. *American Journal of Physiology-Regulatory, Integrative and Comparative Physiology, 293*(5), R1923–R1937.
153. Schneidman, E., Freedman, B., & Segev, I. (1998). Ion channel stochasticity may be critical in determining the reliability and precision of spike timing. *Neural Computation, 10*(7), 1679–1703.
154. Schreiber, T. (1993). Determination of the noise level of chaotic time series. *Physical Review E, 48*(1), R13.
155. Schreiber, T. (1993). Extremely simple nonlinear noise-reduction method. *Physical Review E, 47*(4), 2401.
156. Shanahan, M. (2010). Metastable chimera states in community-structured oscillator networks. *Chaos: An Interdisciplinary Journal of Nonlinear Science, 20*(1), 013108. https://doi.org/10.1063/1.3305451
157. Shipp, S. (2007). Structure and function of the cerebral cortex. *Current Biology, 17*(12), R443–R449.
158. Siefert, M., Kittel, A., Friedrich, R., & Peinke, J. (2003). On a quantitative method to analyze dynamical and measurement noise. *Europhysics Letters, 61*(4), 466.
159. Simpson, D. M., & Wicks, R. (1988). Spectral analysis of heart rate indicates reduced baroreceptor-related heart rate variability in elderly persons. *Journal of Gerontology, 43*(1), M21–M24.
160. Sinski, M., Lewandowski, J., Abramczyk, P., Narkiewicz, K., & Gaciong, Z. (2006). Why study sympathetic nervous system. *Journal of Physiology and Pharmacology, 57*(Suppl 11), 79–92.
161. Song, A. W. (2012). Diffusion modulation of the fMRI signal: Early investigations on the origin of the BOLD signal. *Neuroimage, 62*(2), 949–952.
162. Stacey, N., & D'Amico, S. (2021). Adaptive and dynamically constrained process noise estimation for orbit determination. *IEEE Transactions on Aerospace and Electronic Systems, 57*(5), 2920–2937. Cited by: 15; All Open Access, Bronze Open Access, Green Open Access.
163. Stewart, I. M. (2009). Maximum-likelihood detection of sources among poissonian noise. *Astronomy & Astrophysics, 495*(3), 989–1003.
164. Stoica, P., Moses, R. L., et al. (2005). *Spectral analysis of signals* (Vol. 452). Pearson Prentice Hall.
165. Strogatz, S. (2004). Sync: The emerging science of spontaneous order. Penguin UK.
166. Strumik, M., Macek, W. M., & Redaelli, S. (2005). Discriminating additive from dynamical noise for chaotic time series. *Physical Review E—Statistical, Nonlinear, and Soft Matter Physics, 72*(3), 036219.
167. Sunagawa, K., Kawada, T., & Nakahara, T. (1998). Dynamic nonlinear vago-sympathetic interaction in regulating heart rate. *Heart and Vessels, 13*(4), 157–174.
168. Szendro, P., Vincze, G., & Szasz, A. (2001). Pink-noise behaviour of biosystems. *European Biophysics Journal, 30*, 227–231.
169. Szpiro, G. G. (1993). Measuring dynamical noise in dynamical systems. *Physica D: Nonlinear Phenomena, 65*(3), 289–299.
170. Tahsili-Fahadan, P., & Geocadin, R. G. (2017). Heart–brain axis: Effects of neurologic injury on cardiovascular function. *Circulation Research, 120*(3), 559–572.
171. Takens, F. (1981). Dynamical systems and turbulence. *Warwick, 1980*, 366–381.

172. Tan, J. P. L. (2017). Simple noise-reduction method based on nonlinear forecasting. *Physical Review E, 95*(3), 032218.
173. Tang, Y., Qian, G., Wu, W., & Chien, Y.-R. (2023). An efficient filtering algorithm against impulse noise in communication systems. *Journal of Internet Technology, 24*(2), 357–362.
174. Taylor, J. R., Williams, N., Cusack, R., Auer, T., Shafto, M. A., Dixon, M., Tyler, L. K., Henson, R. N., et al. (2017). The Cambridge Centre for ageing and neuroscience (Cam-CAN) data repository: Structural and functional MRI, MEG, and cognitive data from a cross-sectional adult lifespan sample. *Neuroimage, 144*, 262–269.
175. Tian, L., Hires, S. A., Mao, T., Huber, D., Chiappe, M. E., Chalasani, S. H., Petreanu, L., Akerboom, J., McKinney, S. A., Schreiter, E. R., et al. (2009). Imaging neural activity in worms, flies and mice with improved gcamp calcium indicators. *Nature Methods, 6*(12), 875–881.
176. Tolhurst, D. J., Movshon, J. A., & Dean, A. F. . (1983). The statistical reliability of signals in single neurons in cat and monkey visual cortex. *Vision Research, 23*(8), 775–785.
177. Tsuji, H., et al. (1994). Reduced heart rate variability and mortality risk in an elderly cohort. The framingham heart study. *Circulation, 90*(2), 878–883.
178. Tuzlukov, V. (2018). *Signal processing noise*. CRC Press.
179. Udhayakumar, R. K., Karmakar, C., & Palaniswami, M. (2017). Approximate entropy profile: A novel approach to comprehend irregularity of short-term HRV signal. *Nonlinear Dynamics, 88*, 823–837.
180. Ulichney, R. A. (1988). Dithering with blue noise. *Proceedings of the IEEE, 76*(1), 56–79.
181. Urbanowicz, K., & Hołyst, J. A. (2003). Noise-level estimation of time series using coarse-grained entropy. *Physical Review E, 67*(4), 046218.
182. Vadali, S. R. K., Ray, P., Mula, S., & Varshney, P. K. (2017). Linear detection of a weak signal in additive cauchy noise. *IEEE Transactions on Communications, 65*(3), 1061–1076.
183. Vaillancourt, D. E., & Newell, K. M. (2002). Changing complexity in human behavior and physiology through aging and disease. *Neurobiology of Aging, 23*(1), 1–11.
184. Valentim, C. A., Inacio, C. M. C., Jr, & David, S. A. (2021). Fractal methods and power spectral density as means to explore EEG patterns in patients undertaking mental tasks. *Fractal and Fractional, 5*(4), 225.
185. Valenza, G., Citi, L., Garcia, R. G., Taylor, J. N., Toschi, N., & Barbieri, R. (2017). Complexity variability assessment of nonlinear time-varying cardiovascular control. *Scientific Reports, 7*(1), 1–15.
186. Valenza, G., Citi, L., Saul, J. P., & Barbieri, R. (2018). Measures of sympathetic and parasympathetic autonomic outflow from heartbeat dynamics. *Journal of Applied Physiology, 125*(1), 19–39.
187. Valenza, G., Citi, L., Scilingo, E. P., & Barbieri, R. (2014). Inhomogeneous point-process entropy: An instantaneous measure of complexity in discrete systems. *Physical Review E, 89*(5), 052803.
188. Valenza, G., Passamonti, L., Duggento, A., Toschi, N., & Barbieri, R. (2020). Uncovering complex central autonomic networks at rest: A functional magnetic resonance imaging study on complex cardiovascular oscillations. *Journal of the Royal Society Interface, 17*(164), 20190878.
189. Valenza, G., Pioggia, G., Armato, A., Ferro, M., Scilingo, E. P., & De Rossi, D. (2011). A neuron–astrocyte transistor-like model for neuromorphic dressed neurons. *Neural Networks, 24*(7), 679–685.
190. Valenza, G., Sclocco, R., Duggento, A., Passamonti, L., Napadow, V., Barbieri, R., & Toschi, N. (2019). The central autonomic network at rest: Uncovering functional MRI correlates of time-varying autonomic outflow. *Neuroimage, 197*, 383–390.
191. Van Essen, D. C., Smith, S. M., Barch, D. M., Behrens, T. E. J., Yacoub, E., & Ugurbil, K. (2013). The WU-minn human connectome project: An overview. *Neuroimage, 80*, 62–79.
192. Vogel, H. (2009). *Nervous system*. Cambridge University Press.
193. White, J. A., Rubinstein, J. T., & Kay, A. R. (2000). Channel noise in neurons. *Trends in Neurosciences, 23*(3), 131–137.

194. Xu, P., Li, W. K., & Jayawardena, A. W. (2012). Noise level estimation for a chaotic time series. *International Journal of Bifurcation and Chaos, 22*(03), 1250052.
195. Yao, T.-L., Liu, H.-F., Xu, J.-L., & Li, W.-F. (2012). Noise-level estimation of noisy chaotic time series based on the invariant of the largest lyapunov exponent. *Acta Physica Sinica, 61*, 060503. https://doi.org/10.7498/aps.61.060503
196. Yao, T.-L., Liu, H.-F., Xu, J.-L., & Li, W.-F. (2012). Estimating the largest lyapunov exponent and noise level from chaotic time series. *Chaos: An Interdisciplinary Journal of Nonlinear Science, 22*(3), 033102.
197. Zhang, Z., & Constandinou, T. G. (2023). Firing-rate-modulated spike detection and neural decoding co-design. *Journal of Neural Engineering, 20*(3), 036003.
198. Zyma, I., Tukaev, S., Seleznov, I., Kiyono, K., Popov, A., Chernykh, M., & Shpenkov, O. (2019). Electroencephalograms during mental arithmetic task performance. *Data, 4*(1), 14.

Index

A. Scarciglia et al., *Physiological Noise and Informative Randomness in Neural and Cardiovascular Systems*, https://doi.org/10.1007/978-3-031-95914-1

The manufacturer's authorised representative in the EU is Springer Nature Customer Service Centre GmbH, Europaplatz 3, 69115 Heidelberg, Germany. If you have any concerns regarding our products, please contact ProductSafety@springernature.com

Printed and bound by CPI Group (UK) Ltd, Croydon, CR0 4YY
30/09/2026
02225577-0015